Biorefinery in the Pulp and Paper Industry

Biorefinery in the Pulp and Paper Industry

Pratima Bajpai

AMSTERDAM • BOSTON • HEIDELBERG • LONDON
NEW YORK • OXFORD • PARIS • SAN DIEGO
SAN FRANCISCO • SINGAPORE • SYDNEY • TOKYO
Academic Press is an imprint of Elsevier

ELSEVIER

Academic Press is an imprint of Elsevier
32 Jamestown Road, London NW1 7BY, UK
The Boulevard, Langford Lane, Kidlington, Oxford, OX5 1GB, UK
Radarweg 29, PO Box 211, 1000 AE Amsterdam, The Netherlands
225 Wyman Street, Waltham, MA 02451, USA
525 B Street, Suite 1900, San Diego, CA 92101-4495, USA

First published 2013

Notices
Knowledge and best practice in this field are constantly changing. As new research and
experience broaden our understanding, changes in research methods, professional practices,
or medical treatment may become necessary.

Practitioners and researchers must always rely on their own experience and knowledge in
evaluating and using any information, methods, compounds, or experiments described herein.
In using such information or methods they should be mindful of their own safety and the safety
of others, including parties for whom they have a professional responsibility.

To the fullest extent of the law, neither the Publisher nor the authors, contributors,
or editors, assume any liability for any injury and/or damage to persons or property as
a matter of products liability, negligence or otherwise, or from any use or operation of
any methods, products, instructions, or ideas contained in the material herein.

British Library Cataloguing-in-Publication Data
A catalogue record for this book is available from the British Library

Library of Congress Cataloging-in-Publication Data
A catalog record for this book is available from the Library of Congress

ISBN: 978-0-12-409508-3

For information on all Academic Press publications
visit our website at store.elsevier.com

This book has been manufactured using Print On Demand technology. Each copy is produced
to order and is limited to black ink. The online version of this book will show color figures
where appropriate.

Working together to grow
libraries in developing countries

www.elsevier.com | www.bookaid.org | www.sabre.org

ELSEVIER   **BOOK AID**
International   Sabre Foundation

CONTENTS

PREFACE

The traditional pulp and paper producers are facing new competitors in tropical and subtropical regions who use the latest and largest installed technologies, and also have wood and labor cost advantages. Due to the increasing global competition, the forest products prices will continue to decrease. To remain viable, the traditional producers need to increase revenue by producing bioenergy and biomaterials in addition to wood, pulp, and paper products. In this so-called "integrated products biorefinery," all product lines are highly integrated and energy efficient. Integrated products biorefineries present the forest products industry with a unique opportunity to increase revenues and improve environmental sustainability. Integrated products biorefinery technologies will allow industry to manufacture high-value chemicals, fuels, and/or electric power, while continuing to produce traditional wood, pulp, and paper products. The industry already controls much of the raw material and infrastructure necessary to create integrated products biorefineries, and Agenda 2020 partnerships are speeding development of the key-enabling technologies. Once fully developed and commercialized, these technologies will produce enormous energy and environmental benefits for the industry and the nation. This e-book presents biorefining concepts, the opportunities for the pulp and paper industry, describes and discusses emerging biorefinery process options, and highlights the environmental impact and the complex and ambiguous decision-making challenges that mills considering implementing the biorefinery will face.

ABBREVIATIONS

2,3-BD	2,3-butanediol
ABE	acetone butanol ethanol
BL	black liquor
BLG	black liquor gasification
BLS	black liquor solids
CC	combined cycle
CCC	counter current condenser
CFD	computational fluid dynamics
DOE	Department of Energy
DME	dimethylether
EA	effective alkali
EvGT	evaporative gas turbine
FAEs	feruloyl esterases
GTL	gas-to-liquids
HAT	humid air turbine
HFCS	high-fructose corn syrup
HRSG	heat recovery steam generator
IEA	International energy efficiency
IER	ion-exchange resin
IFPB	integrated forest products biorefinery
IGCC	integrated gasification and combined cycle
MTBE	methyltertiarybutylether
MTCI	manufacturing and technology conversion International
MWCO	molecular weight cutoff
NO_x	nitrogen oxides
PEO	polyethylene oxide
PHB	poly-3-hydroxybutyric acid
STIG	steam-injected gas turbine
TRI	ThermoChem Recovery International
TRS	total reduced sulfur
VMD	vacuum membrane distillation
VOCs	volatile organic compounds

Biorefinery Concept*

The reliability, affordability, and environmental impact of energy supplies have become the most critical issue for the world economy. Due to world population growth, primary energy consumption has increased and will continue to increase in the future. This energy pool is mostly fossil-carbon-based and is predominantly used for transportation and energy production purposes (Sousa, 2010). According to the International Energy Agency (IEA) about 50% of this energy demand is for transportation purposes only (IEA, 2007). As a result of this, oil price has increased, and has also affected other primary energy sources prices (BP Statistical Review of World Energy, 2009). This situation, along with the need for reducing foreign oil dependency and the environmental awareness of world's population has led to a search for alternative primary energy and carbon sources based upon renewable sources. Biomass, especially lignocellulosic material, represents an abundant renewable carbon source. This is potentially convertible to energy, fuels, and speciality chemicals. The integrated production of bioenergy, biofuels, and biochemicals, through advanced technological processes of separation and conversion that minimizes carbon cycle impact, defines the biorefinery concept. The term "biorefinery" is a refinery utilizing forestry and agricultural biomass as a feedstock to produce gaseous and liquid fuels, specialty or commodity chemicals, or other products commonly produced in petrochemical refineries, where the feedstocks are mainly fossil fuels (Axegård, 2005, 2007). Figure 1.1 shows integrated forest industry biorefinery (Hetemäki, 2007).

*Some excerpts taken from Bajpai (2012). Biotechnology for Pulp and Paper Processing with kind permission from Springer Science + Business Media.

Biorefinery in the Pulp and Paper Industry. DOI: http://dx.doi.org/10.1016/B978-0-12-409508-3.00001-8

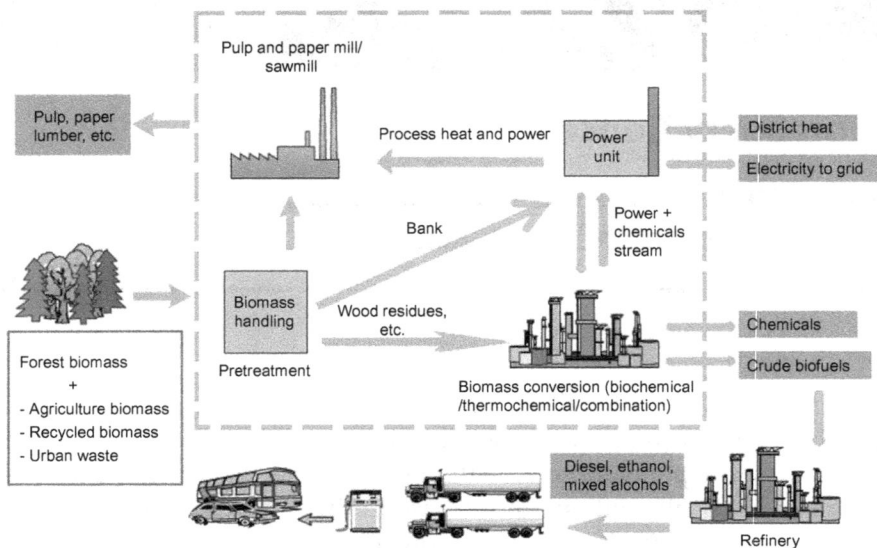

Figure 1.1 Integrated forest industry biorefinery. Reproduced with permission from Hetemäki (2007).

The topic of biorefineries as a means of processing industrial material and efficient utilization of renewable products is well known and applied worldwide, in almost every developed and emerging country. The forest products industry's manufacturing facilities are an ideal foundation to develop the Integrated Forest Products Biorefinery (IFPB). Those facilities, which today produce pulp, paper, and wood products, also are geared to collect and process biomass. Rather than creating a "greenfield" operation, additional bioconversion or thermochemical processes can be built around existing mills (either as extensions of the mill or as "across-the-fence" operations) to generate bioenergy or manufacture bioproducts. This presents industry with a dramatic potential to increase the productivity and profitability of its manufacturing infrastructure. Possible benefits include: improved efficiency of raw material utilization, protection of traditional product lines, creation of higher skilled and better paying jobs, and access to new domestic and international markets for bioenergy and bioproducts.

The conversion of forestry and agricultural biomass is accomplished by various extraction and transformation pathways, offering the opportunity to revitalize the pulp and paper industry (Towers et al., 2007). The main biorefinery feedstocks are hemicellulose, cellulose, lignin, and bark, used to generate building block molecules, chemicals,

Table 1.1 Drivers for Pulp Mill Biorefining

- Economic pressures of pulp production
- Reduce dependence on petroleum
- Improve profits of the stagnant paper industry
- Competition for biomass from the energy sector
- Processing of large volumes of biomass
- Infrastructure in place
- Global incentives for fuels/chemicals from biomass (incentives, taxes, credits)
- New efficient separation processes available

fuels, polymers, or dissolving pulp. A large-scale implementation of the biorefinery will result in profitable and sustainable processes with positive environmental impacts (Marinova et al., 2010).

In recent years, there has been great interest in the concept of the forest biorefinery from the forest industry, research community and policy-makers (Bajpai, 2012; Balensiefer, 2008; Bush, 2006; Cunningham, 2005; Johnson et al., 2009; Magdzinski, 2006; Mateos-Espejel et al., 2011; Montréal Workshop, 2005; Ragaukas, 2006; Realff and Abbas, 2004; Thorp, 2005a,b; Thorp and Raymond, 2005; Thorp et al., 2008; Towers et al., 2007; Yunqiao et al., 2008). The concept is attractive because it addresses current concerns of oil prices, finite fossil resources, and Kyoto commitments. Biomass-rich nations see an opportunity to utilize their natural bioresources in new ways to achieve maximum value and productivity within the confines of sustainability. Table 1.1 shows some drivers for pulp mill biorefining.

Some of today's pulp and paper mills are already operating as rudimentary forest biorefineries. Byproducts from the pulping process are used in boilers to produce heat and power, and in some cases, marketable products such as kerosene, tall oil, and cellulose derivatives are generated in addition to paper products. In the optimized forest biorefinery, advanced technologies would enable more of the wood feedstock to be converted to higher-valued products, including chemicals and more marketable fuels such as ethanol and hydrogen.

The products produced in a biorefinery will mainly be a function of the feedstocks available (Towers, 2007). Properties of feedstock such as cost, location, composition, moisture content, and availability will determine the appropriate technical options. Feedstock costs can represent a large portion of plant operating costs. One approach is to locate the biorefinery near the feedstock to reduce or eliminate transportation costs. If the feedstock is a waste stream from an existing process, disposal, or treatment costs may be counterbalanced, resulting in a near-zero feedstock cost. About half the organic mass that enters a kraft mill is incinerated. Capturing more useful energy from this organic mass is the main goal of mill energy efficiency programs. Boiler temperatures and pressures have increased over time to increase power production in steam turbines. However, conventional cogeneration does not represent the ceiling of usefulness for these feedstocks; rather, it represents the floor. In addition to the recovery boiler, most kraft mills operate a wood waste boiler. Most require this additional steam to operate their process, but there is significant opportunity to improve process energy efficiency to eliminate any fossil fuel used to generate steam and to liberate feedstocks for biorefinery opportunities.

In addition to available on-site feedstocks, conventional forestry practices leave residuals—branches, foliage and tree tops—on the forest floor. These represent 15–20% of the tree mass above the root and are generally not utilized. In some countries, residuals are collected and used as fuel for large combined heat and power plants. Tax incentives meant to reduce fossil fuel use in response to Kyoto have been a driving force behind these practices. In some countries, wealth of natural resources and inexpensive hydroelectric power have been the main barriers to implementing similar practices. However, this is changing, as demand in regions normally in surplus of hydroelectric power is growing beyond the installed capacity, leading to increased reliance on high cost incremental capacity. Pulp and paper mills are typically the largest industrial infrastructure located near forestry residuals. Despite that, transportation to the mill site has been highlighted as an obstacle to utilizing this material. Mobile energy densification technologies have been suggested. These technologies are truck-based pyrolysis units that concentrate these residuals for transport to a utilization site. It may also be attractive to utilize nonforestry biomass waste in some cases. According to Rooks (2006), agricultural residuals or even energy crops may provide interesting opportunities.

There is no singular concept of a biorefinery. In fact, there is an infinite amount of concepts depending on the combination of production pathways from the raw material towards the finished products.

Sousa (2010) reported some of the possible biorefinery production platforms that can be adopted in a biorefinery. Of all the platforms considered, the conversion platform is the one that presents more challenges. Several new processes are being developed to answer these new challenges. The main processes are associated to two different technology bases—the biochemical platform and the thermochemical platform (Table 1.2).

1.1 BIOCHEMICAL PLATFORM

A biochemical platform is an application of several extraction, separation, and biological conversion processes of biomass elemental components in order to produce biofuels and biochemicals. The sugar-to-ethanol processes, lignin valorization, and vegetable sterols extraction are examples of different technological routes within this platform.

Table 1.2 Biorefinery's Main Production Platforms

Platform	Raw Materials	Processes	Products	Status
Biochemical	Lignocellulosic and starch biomass	Chemical and enzymatic hydrolysis, fermentation, biotransformation, chemical and catalytical processes	Added value chemicals (both from sugar and lignin), building block chemicals, materials (from lignin or lignocellulose), fuel ethanol, heat and electricity (from lignin)	Laboratory, large scale pilot plant and commercial fuel ethanol (sugarcane, starch-based)
Thermochemical	Lignocellulosic biomass but also plastics and rubber	Thermochemical processes − Gasification − Pyrolysis	Syngas, pyrolysis oil, added value chemicals, gaseous or liquid fuels	Laboratory, large scale pilot plant
Biogas	Liquid effluents, manure	Anaerobic digestion	Methane and CO_2 (biogas), added value chemicals	Commercial Large scale pilot plant
Carbon-rich chains (oil)	Plant oils such as soybean, rapeseed (biodiesel), corn, palm, canola oils, animal fat	Transesterification	Fatty acid methyl ester (biodiesel), glycerin and fatty acids as platform chemicals	Commercial

Source: Based on Carvalheiro et al. (2008).

Wood and woody residues are mainly constituted by sugars, which can ultimately be biologically converted to biofuels or bioproducts. However, sugars are involved in a complex lignocellulosic matrix, which hampers their extraction and consequently their conversion. In addition, one of the main sugars present on biomass is a C5 sugar, like xylan, which is harder to ferment than the C6 sugars, like glucose. Thus, there are several challenges to overcome in order to deploy this technology, namely: biomass pretreatment stage, sugar hydrolysis, and C5 sugar fermentation.

The pretreatment stage can be adjusted according to the specific objective:

a. full conversion of biomass
b. preextraction of wood sugars prior to kraft cooking.

Full conversion of biomass implies that pretreatment stage is directed to increasing the accessibility of all carbohydrates within the crystalline structure of cellulose. In the second case, a mild pretreatment is required in order to selectively extract hemicelluloses from wood. These sugars are degraded inevitably in kraft cooking and are used as a low calorific value fuel in black liquor. Thus, wood hemicelluloses preextraction and its conversion to ethanol, or other products, should add more value to the kraft process. The main biomass pretreatment processes include autohydrolysis, dilute acid hydrolysis, and steam explosion. After biomass pretreatment, the solid fraction has to be hydrolyzed and fermented. There are several process alternatives, with increasing integration and cost reduction potential but also with increasing complexity:

1. Separated hydrolysis and fermentation
2. Simultaneous sacharification and fermentation
3. Simultaneous sacharification and cofermentation
4. Direct microbial conversion.

Each one of these alternatives differs on the integration of the sugar hydrolysis, pentose fermentation, and hexose fermentation stages.

Top value-added chemicals that can be produced in a biochemical platform biorefinery have been identified (Werpy and Petersen, 2004) (Table 1.3).

Table 1.3 Top Value-Added Chemicals Produced in a Biochemical Platform Biorefinery			
Amino Acids	**Carboxylic acid**	**Polyols**	**Lactones**
Aspartic acid	Succinic acid	Glycerol	3-Hydroxy butyrolactone
Glutamic acid	Fumaric acid	Arabitol	
	Malic acid	Xylitol	
	Itaconic acid	Sorbitol	
	Levulinic acid		
	Glucaric acid		
	2,5 Furan dicarboxylic acid		
Source: *Based on Carvalheiro et al. (2008).*			

1.2 THERMOCHEMICAL PLATFORM

The thermochemical platform can be defined as group of biomass thermal treatment processes that uses the production of syngas or biooil as a building block to their conversion in bioenergy, biofuels, and biochemicals. Biomass gasification, black liquor gasification, biomass pyrolysis or liquefaction, and carbonisation, oils aqueous phase reforming are examples of different technological routes within this platform (Katofsky et al., 2003; Wising and Stuart, 2006).

Biomass, either solid, as in the case of wood residues, or liquid, as in the case of black liquor, can be converted through a thermal process to a synthetic gas comprised mainly of CO and H_2 with low amounts of methane, CO_2, and H_2O (Towers, 2007). There are different system configurations depending on the type of reactor, pressure level, gasification agent, and type of heating. The tar elimination and/or cleaning specifications of syngas are the main technological barriers to overcome. After solid biomass and the residual pulping liquors have been gasified, there is a choice of turning the synthesis gas into power or into liquid fuels and/or chemicals.

The gasification system can be coupled with a combined cycle power production unit. Higher power production efficiencies of 30−40% are expected with this system along with lower atmospheric emissions. High capital costs and large-scale operation inexperience hinder its commercial application. Technical and economical viability should be assessed within this option. The use of an internal combustion engine is another option for a small-scale application. Normally, diesel and gas are used within this technology. Landfill gases or methane that are biologically generated have also been used. The main problem concerns the gas cleaning solution. The power production efficiency of these systems can reach up to 20–30%. Finally, burning syngas in dedicated boilers or cofiring it in coal, fuel, and recovery boilers is the simplest application of biomass gasification product. In pulp and paper industry, it can also be employed in limekiln operation, replacing fossil fuels. Syngas can also be upgraded to produce synthetic natural gas improving thermal gas applications.

The combinations of raw materials, conversion process/technology, and final products portfolio, associated to any of these platforms are virtually unlimited. Moreover, there is no defined border between these two technology concepts, as there could be many synergetic interactions that can maximize environmental and economical product value. The final decision about biofuels, biochemicals, and bioenergy production pathway will depend upon raw material availability, technological know-how, regional policies, market regulations, and dynamics (Sousa, 2010).

REFERENCES

Axegård, P., 2005. The future pulp mill—a biorefinery. In: First International Biorefinery Workshop. Washington, DC, July 20–21, 2005.

Axegård, P., 2007. The kraft pulp mill as a biorefinery. Third ICEP International Colloquium on Eucalyptus Pulp. 4–7 March, Belo Horizonte, Brazil, p. 6.

Bajpai, P., 2012. Biotechnology in Pulp and Paper Processing. Springer-Verlag Inc., New York, NY.

Balensiefer, T., 2008. Chemicals from biomass: an industrial perspective. Biorefining for the Pulp and Paper Industry Conference. 3 December 2008, 18–19 September, Helsinki.

BP, 2009. BP Statistical Review of World Energy, June 2009.

Bush, G.W., 2006. State of the Union Address, January 31.

Carvalheiro, F., Duarte, L.C., Gírio, F.M., 2008. Hemicellulose biorefineries: a review on biomass pretreatments. J. Sci. Ind. Res. 67, 849–864.

Cunningham, J.E., 2005. Biomass as an industrial feedstock. Canadian Chem. News 24–26 June.

Hetemäki, L., 2007. Forest biorefineries: current status and outlook. IUFRO Division VI Symposium. 14—20 August 2007, Saariselkä, Finland.

International Energy Association (IEA), 2007. World Energy Outlook 2007: China and India Insights. IEA Publications, Paris, France.

Johnson, B., Johnson, T., Scott-Kerr, C., Reed, J., 2009. The Future Is Bright. Pulp Paper Int. October, 19–22.

Katofsky, R., Consonni, S., Larson, E.D., 2003. A cost–benefit analysis of black liquor gasification combined cycle systems. Proceedings, TAPPI Fall Technical Conference: Engineering, Pulping and PCE&I. Chicago. p. 22.

Magdzinski, L., 2006. Bioenergy and bioproducts at Tembec: synergies in integrated processing of biomaterials. PAPTAC Ninety-Second Annual Meeting. Montreal.

Marinova, M., Mateos-Espejel, E., Paris, J., 2010. From kraft mill to forest biorefinery: an energy and water perspective II. Case study. Cellul. Chem. Technol. 44 (1–3), 21–26.

Mateos-Espejel, E., Moshkelani, M., Keshtkar, M., Paris, J., 2011. Sustainability of the green integrated forest biorefinery: a question of energy. J. Sci. Technol. For. Prod. Processes 1 (1), 55.

Montréal Workshop on Bio-refineries, 2005. Capturing Canada's Natural Advantage. November 21, Montréal QC.

Ragaukas, A.J., 2006. The path forward for biofuels and biomaterials. Science 311 (1), 484–489.

Realff, M.J., Abbas, C., 2004. Industrial symbiosis: refining the biorefinery. J. Ind. Ecol. 7 (3–4), 5–9.

Rooks, A., 2006. A sweet future for bio-fuels? Solutions 89 (2), 4.

Sousa, G.D.A., 2010. Development pathways: a survey for the pulp and paper industry. XXI Encontro Nacional da TECNICELPA/ VI CIADICYP 2010, 12–15 Outubro 2010, Lisboa, Portugal.

Thorp, B., 2005a. Transition of mills to biorefinery model creates new profit streams. Pulp Pap. November, 35–39.

Thorp, B., 2005b. Biorefinery offers industry leaders business model for major change. Pulp Pap. 79 (11), 35–39.

Thorp, B., Raymond, D., 2005. Forest biorefinery could open door to bright future for P&P industry. PaperAge 120 (7), 16–18.

Thorp, B.A., Thorp, B.A., Murdock-Thorp, L.D., 2008. A compelling case for integrated biorefineries <http://www.epoverviews.com/oca/Compellingcaseforbiorefineries.pdf> (accessed 15.11.2012).

Towers, M., Browne, T., Kerekes, R., Paris, J., Tran, H., 2007. Biorefinery opportunities for the Canadian pulp and paper industry. Pulp Pap. Canada 108 (6), 26–29.

Werpy, T., Petersen, G., 2004. Top value-added chemicals from biomass, Volume I: results of screening for potential candidates from sugars and synthesis gas, pacific northproduct west National Laboratory, August 2004 <http://www.eere.energy.gov/biomass/pdfs/35523.pdf>.

Wising, U., Stuart, P.R., 2006. Identifying the Canadian forest biorefinery. Pulp Pap. Canada 107 (6), 25–30.

Yunqiao, P., Zhang, D., Singh, P., Ragauskas, A., 2008. The new forestry biofuels sector. Biofuels Bioprod. Biorefin. 2 (1), 58–73.

CHAPTER 2

Biorefinery Opportunities in the Pulp and Paper Industry*

2.1 BIOREFINERY OPPORTUNITIES

Pulp and paper mills are ideal sites for biorefineries (Connor, 2007). The reasons are: paper and forest products companies are efficient growers, harvesters, transporters, and processors of biomass; pulp and paper mills are located near numerous sources of biomass, such as forest and agricultural residuals, and energy crops, and have existing infrastructure to ship finished product; paper mills utilize over several million dry tons of wood per year as a raw material. Additionally, these mills have ready access to a roughly equal amount of forest residuals and an even greater amount of agricultural waste and energy crops; and pulp and paper mills are familiar with producing power from biomass — currently they produce 60% of their power from wood residuals and spent liquors. Pulp and paper mills also have a highly trained workforce capable of operating energy and biorefinery systems. Pulp and paper companies are may consider following one of the general biorefinery configurations (Chambost et al., 2008):

- In the simplest case, biorefinery configurations that try to transform biomass into biofuels replacing bunker carbon or natural gas currently consumed in the pulp and paper manufacturing processes, and potentially benefitting from an improved carbon footprint and carbon credits.
- Greenfield biorefineries, versus those implemented in retrofit to operating pulp and paper mills, versus brownfield biorefineries that are developed around pulp and paper mill closures.
- Biorefineries that try to capitalize on the emerging huge biofuels market via ethanol or diesel production, versus biorefineries seeking to focus

*Some excerpts taken from Bajpai (2012). Biotechnology for Pulp and Paper Processing with kind permission from Springer Science + Business Media.

Biorefinery in the Pulp and Paper Industry. DOI: http://dx.doi.org/10.1016/B978-0-12-409508-3.00002-X

more on smaller volume bioproducts with potentially having a higher added value.
- And in the most complex, and potentially most value-generating configuration for the forest products industry, is to try to establish a fully integrated forest biorefinery along with the proven and needed pulp, paper, and paperboard manufacturing operations. In this highly evolved platform mode, the aggregate benefits of operating and management synergies will emerge across time.

Many investigators are assessing biorefinery technologies and pathways (Chambost et al., 2007, 2008; Janssen et al., 2008; Stuart, 2006). However, these analyses are typically limited to technoeconomics and lack any analyses of product design and new customer linkages. Forest biorefinery design must consider a multidimensional and multidisciplinary approach, including customer/product and process concepts in order to make the biorefinery not only a capital project, but an initiative that results in a successful business transformation. A systematic methodology for helping industry executives decide which technologies to implement, and which biorefinery products to manufacture, is needed.

By integrating forest biorefinery activities at an existing plant, pulp and paper mills have the opportunity to produce significant amounts of bioenergy and bioproducts and to drastically increase their revenues while continuing to produce wood, pulp, and paper products. Manufacturing new value-added byproducts (e.g. biofuels, bulk and specialty chemicals, and pharmaceuticals) from biomass represents an unprecedented opportunity for revenue diversification for some forestry companies. The biorefinery builds on the same principles as the petrochemical refinery. In a petrochemical refinery, the raw material is normally crude oil, whereas in the forest biorefinery the raw material is wood/biomass. The raw material stream is fractionated into several product streams. The products can be a final product or a raw material for another process. New technology is being developed that could be integrated into an existing pulp and paper mill, transforming it into a forest biorefinery. There are still significant challenges associated with these new technologies, but several of them look promising. Research is emerging that is focused on biorefinery technology development in North America and around the world (Bajpai, 2012; Closset, 2004; Mabee et al., 2005). However, these process technology development activities alone do not address most of the significant risks associated with implementing the forest biorefinery.

Biorefinery technology development will typically be implemented as a retrofit and must be accompanied by a careful process systems analysis in order to understand the impact on existing processes, e.g., pulp yield reductions because carbon is used to make alternative products, and the potential for changed black liquor scaling characteristics in evaporators. The objective of this process systems analysis is to preserve the value of the existing pulp and paper-producing asset.

In addition to process technology development, product development will be essential for identifying successful new markets for biorefinery products, and their supply chain management strategies. These are again systems-oriented issues where the evaluation will be critical for the success of the forest biorefinery.

The current pulp and paper mills use logs and fiber, chemicals and energy to produce commodity pulp and paper products (Connor, 2007; Thorp et al., 2008). Future mills, which we can consider Integrated Forest Biorefineries, will import regional biomass instead of purchased energy. They will expand the industry's mission from simply manufacturing low-margin paper products to creating new revenue streams by producing "green" power and creating new, high-value products such as biofuels and biochemicals, all while improving the efficiency and profitability of their core paper-making operations.

Kraft pulp mills are the primary candidates to be transformed into biorefineries, as they have the infrastructure to process biomass feedstock (Hytonen and Stuart, 2009). The retrofit integration of a biorefining technology into a kraft process will add new operations and will increase the steam demand; also, if hemicellulose or lignin is removed, it will reduce the fuel available to the recovery boiler (Marinova et al., 2010). It is important that the base process, as well as the biorefinery unit, is optimized and highly integrated from the standpoint of energy, to satisfy the modified energy balance.

Table 2.1 shows some examples of forest industry biorefinery products. The products developed within biorefineries may offer many possibilities for reducing greenhouse gases and countering the effects of declining fossil fuel availability. In addition, biorefinery products may provide economic benefits through the replacement of natural gas and superior utilization of agricultural and other biological waste (Covey and Grist, 2008). Renewable energy is growing from a very small base

Table 2.1 Forest Industry Biorefinery Products: Some Examples	
Intermediate products	Important features
• Chips	• Multioutput technology
• Sawdust	• Products can be complements or
• Lignin	substitutes
• Sugars	• Intermediate products can play a central
• Lipids (Fatty and rosin acids)	role for the economic viability
Final products	• Potential for CO_2 reduction through product substitution
• Pulp, paper, boards	
• Wood products	
• Heat	
• Electricity	
• Ethanol, diesel, DME, hydrogen	
• Polymers	

and examples of the difficulties of applying current renewable energy technology to large-scale applications have been demonstrated in both wind power and solar energy. Whilst over half of the world's oil is used to make transport fuels, in the near future such fuels are likely to be partly replaced with ethanol or other liquid chemicals produced from sugars and polysaccharides. The technology used in the development of these fuel substitutes is much more familiar to the pulp industry than the oil industry and there is scope for growth in the pulp industry by serving this market. The next largest use for oil is as a fuel for stationary units and, with the current stage of development of synthetic fuels, it is considerably more economical to fire such units with solid biomass. As oil becomes scarce and increasingly expensive, biomass can be expected to replace it in some of the "easier" applications such as ethanol. The increasing interest in biorefining should be taken as an opportunity rather than a threat by the pulp and paper industry.

REFERENCES

Bajpai, P., 2012. Biotechnology in Pulp and Paper Processing. Springer-Verlag Inc., New York, NY.

Chambost, V., Eamer, B., Stuart, P.R., 2007. Forest biorefinery: getting on with the job. Pulp Pap. Canada 108 (2), 19–20 (22).

Chambost, V., Mcnutt, J., Stuart, P.R., 2008. Guided tour: implementing the forest biorefinery (FBR) at existing pulp and paper mills. Pulp Pap. Canada 109 (7-8), 19–27.

Closset, G., 2004. Advancing the forest biorefinery. Presentation at Forest Products Techno-Business Forum. 26–27 October, Atlanta, GA.

Connor, E., 2007. The integrated forest biorefinery: the pathway to our bio-future. International Chemical Recovery Conference: Efficiency and Energy Management. 29 May–1 June, 2007, Quebec City, QC, Canada, pp. 323–327.

Covey, G., Grist, S., 2008. What is the role for biorefineries? Sixty-Second Appita Annual Conference and Exhibition, Rotorua, New Zealand, 20–23 April 2008, Conference Technical Papers. Appita Inc., Carlton, Vic. (1–18).

Hytonen, E., Stuart, P.R., 2009. Integrated bioethanol production into an integrated kraft pulp and paper mill: techno-economic assessment. Pulp Pap. Canada 110 (5), 25–32.

Janssen, M., Chambost, V., Stuart, P.R., 2008. Successful partnerships for the forest biorefinery. Ind. Biotechnol. 4 (4), 352–362.

Mabee, W.E., Gregg, D.J., Saddler, J.N., 2005. Assessing the emerging biorefinery sector in Canada. Appl. Biochem. Biotechnol. 121–124, 765–777.

Marinova, M., Mateos-Espejel, E., Paris, J., 2010. From kraft mill to forest biorefinery: an energy and water perspective 11. Case study. Cellul. Chem. Technol. 44 (1–3), 21–26.

Stuart, P.R., 2006. The forest biorefinery: survival strategy for Canada's pulp and paper sector? Pulp Pap. Canada 107 (6), 13–16.

Thorp, B.A., Thorp, B.A., Murdock-Thorp, L.D., 2008. A compelling case for integrated biorefineries <http://www.epoverviews.com/oca/Compellingcaseforbiorefineries.pdf>.

CHAPTER 3

Emerging Biorefinery Process Options*

Several options should be considered for the implementation of a biorefinery in a pulp and paper mill. The biorefinery technologies currently under development are typically characterized as biochemical and thermochemical processes. The biochemical processes uses steam, dilute acid, concentrated acid, and/or enzyme hydrolysis to convert the hemicellulose and cellulose of biomass into simpler pentoses and glucose. The thermochemical processes use slow or medium temperature gasification or higher temperature pyrolysis to create a high hydrogen content synthetic gas (syngas) that can be used for electricity generation or catalytically converted into liquid biofuels.

The emerging biorefinery process options are shown in Table 3.1 (Wising and Stuart, 2006). Hemicellulose preextraction, lignin precipitation, tall oil extraction technologies are biochemical, and black liquor gasification (BLG) is thermochemical. Hemicellulose

*Some excerpts taken from Bajpai P (2012). Biotechnology for Pulp and Paper Processing with kind permission from Springer Science + Business Media.

Biorefinery in the Pulp and Paper Industry. DOI: http://dx.doi.org/10.1016/B978-0-12-409508-3.00003-1

Table 3.1 Emerging Biorefinery Process Options

- Hemicellulose preextraction
- Black liquor gasification
- Removal of lignin from black liquor
- Tall oil extraction

Source: *Based on Wising and Stuart (2006)*.

preextraction is the most extensively investigated concept of the biorefinery platform. The choice of biorefinery technology will depend firstly on the choice of appropriate products as they relate to markets and the supply chain. Depending on the choice of technologies implemented, the yield, the impact on the pulp and paper process, and the capital cost will vary. As the processes in a pulp and paper mill are strongly linked, it is difficult to foresee the impact implementing these different technologies might have on the entire mill. Plus, adding two or more technologies to one mill brings process issues that are complex to consider. One of the key criteria for forest biorefinery options is that the processes are adaptable (Farmer, 2005). Many of the products that could be produced in a forest biorefinery follow different value cycles. If these products could be changed, the most profitable product could be produced at a time where the value of said product is the highest. By developing a concept of adaptable forest biorefinery, the mill would be less economically vulnerable, as the product produced could change over time.

3.1 PREEXTRACTION OF HEMICELLULOSE

Lignocellulosic materials mainly consists of cellulose, hemicellulose, and lignin. Table 3.2 shows the composition of hardwoods and softwoods, and Table 3.3 shows hemicellulose composition of some hardwoods and softwoods. In hardwoods, the major hemicellulose component is the *O*-acetyl-4-*O*-methylglucuronoxylan, whereas in the softwood species, the *O*-acetylgalactoglucomannan is the predominant component. The building blocks of hemicelluloses are hexoses (glucose, mannose, and galactose) and pentoses (xylose and arabinose) which exist in pyranose and furansose forms. The removal and recovery of hemicellulose is a main step of pretreatment processes for biological conversion to ethanol or other products. A variety of pretreatment

Table 3.2 Chemical Composition of Hardwoods and Softwoods

Wood Components	Hardwoods	Softwoods
Cellulose	40–50	45–50
Hemicellulose	22–35	20–30
Lignin	20–30	25–35
Extractives	1–5	3–8

Table 3.3 Hemicellulose Composition of Some Hardwoods and Softwoods

Substrate	Xylose	Arabinose	Galactose	Mannose	Rhamanose	Uronic Acid
Hardwoods						
Eucalyptus	14–190	0.6–1	1–1.9	1–2	0.3–1	2
Birch	19–25	0.3–0.5	0.7–1.3	1.8–3.	0.6	4–7
Aspen	18–28	0.7–4.0	0.6–2	0.9–2.5	0.5	5–6
Softwoods						
Pine	5–10	2–4	2–4	6–13	–	3–6
Spruce	5–10	1.2	1.9–4.3	9.4–15	0.3	1.5–6

Source: *Based on Menon et al. (2010).*

methods (Brownell and Saddler, 1987; Bozell et al., 1995; Cara et al., 2006; Cunningham et al., 1986; Eckert et al., 2000, 2004; Fitzpatrick, 1997; Gabrielii, et al., 2000; Hashimoto and Hashimoto, 1975; Heitz et al., 1986; Heitz et al., 1991; Hsu, 1996; Huang et al., 2008; Ibrahim and Glasser, 1999; Josefsson et al., 2002; Knappert and Grethlein, 1981; Lazzaroni et al., 2005; Lee et al., 1999; Lesutis et al., 2001; Li et al., 2004; Lu et al., 2004; Moens and Khan, 2003; Mok and Antal, 1992; Mosier et al., 2005; N'Diaye et al., 1996; Nguyen et al., 2000; Nolen et al., 2003; Saha, 2003; Saska and Ozer, 1995; Scott, 1989; Shimizu et al., 1998; Sun et al., 2001; Swatloski et al., 2002; Tucker et al., 2003; Wai et al., 2003; Weil et al., 1997, 1998; Wyatt et al., 2005) to hydrolyze and fractionate hemicellulose components have been studied in detail (Table 3.4).

Pretreatments of lignocellulosic materials by water or steam are referred to in the literature as autohydrolysis (Lora and Wayman, 1978), hydrothermolysis or hydrothermal pretreatment (Kubikova et al., 1996), aqueous liquefaction or extraction (Heitz et al., 1986).

There are several applications of hemicelluloses such as biopolymers, hydrogels or thermoplastic xylan derivatives or, source of sugars

Table 3.4 Methods Used for Hydrolysis of Hemicelluloses Components

- Dilute acid
- Liquid hot water extraction
- Dilute acid–steam explosion
- Alkaline extraction
- Ammonia fiber/freeze explosion (AFEX)
- Organosolv fractionation
- Supercritical carbon dioxide
- Ionic liquids (new class of solvents with nonmolecular, ionic character that are liquids at room temperature)

Source: *Based on Yunqiao et al. (2008); Huang et al. (2008).*

for fermentation to fuels, such as ethanol, or chemicals, such as 1,2,4-butanetriol, a less hazardous alternative to nitroglycerine (Ebringerova et al., 1994; Gabrielii, 2000; Jain et al., 2000; Niu et al., 2003) The cosmetics industry uses hemicelluloses as emulsifiers to prepare water and oil emulsions. Research has also been carried out into hemicelluloses as immunomodulators or those properties that fight infections. The building blocks of hemicelluloses also include sugars with interesting physiological effects. One example of such a sugar is mannose, which has been shown to help combat certain stomach infections. These monosaccharides are currently being studied for example converting xylose into xylitol and mannose into mannitol. These sugars are packed with potential. If hemicelluloses are broken down into smaller pieces or so-called oligomers, there is evidence that these pieces are highly bioactive. There is also data that they promote tree growth or function as growth hormones. Hemicellulose can also be used as a dietary fiber. Their sugars are so-called slow carbohydrates, which help balance blood sugar levels and promote weight loss. Table 3.5 presents use of hemicelluloses as papermaking additives, and Table 3.6 presents some of the most important current and potential applications of xylan in the pulp and paper, pharmaceutical, chemical, food, and fermentation industries.

Not much work has been done on extracting and utilizing hemicelluloses prior to the pulping process. Removal of hemicelluloses from wood as a pretreatment step is presently being practiced commercially in the production of dissolving pulp which is a high purity specialty grade pulp made for processing into cellulose derivatives including cellulose nitrate, cellulose xanthate (rayon fibers), and cellulose acetate (Mateos-Espejel et al., 2011). While demand for this type of pulp has recently increased, current world production capacity cannot meet

Table 3.5 Use of Hemicelluloses as Papermaking Additives
Dry strength

Positive effect of spruce hemicelluloses on mechanical properties
    however moderate compared to cationic starch
Little effect of birch hemicelluloses
    But neutralization of anionic charge of birch hemicelluloses (carried by the xylan fraction)
    − Improves adsorption on pulp
    − Contributes to development of mechanical properties
Bleaching required for white grades

Antihornification

Spruce hemicelluloses
    − improves mechanical properties at first addition (good dry strength aid) but does not maintain
        mechanical properties after multiple drying cycles (not so good antihornification aid)
Birch hemicellulose
    − helps maintain mechanical properties over multiple recycling

Surface sizing (OCC)

Relevant size uptake achieved with all additives
Interesting development of mechanical properties when replacing starch with hemicelluloses
Birch hemicelluloses
    Gains of Burst index (+ 10%), breaking length (+ 5%), CMT_A index (+ 5−10%)
    Brownish color brought by birch hemicellulose is close to desired "kraft" color for testliner
Tear CD index slightly decreases with all additives

Source: *Based on Perez et al. (2011).*

market requirements (high revenues available) (Thurso Project, 2010). The transformation of a kraft pulp mill into a dissolving pulp mill requires the extraction of the hemicellulose prior to pulping. However, hemicellulose is typically sent to the recovery boilers for steam production.

A study to investigate the integration of a hemicellulose-based biorefinery into a Canadian dissolving pulp mill with a production of 500 t/d, has been performed (Marinova et al., 2010a). The amount of hemicellulose extracted in the prehydrolysis step of a dissolving pulp mill varies according to the hydrolysis method—steam, hot water, used and the type of wood. A typical value of 30% of hemicelluloses extracted from woodchips was used in this study. The mill produced 700 t/d of hemicellulose hydrolysate. In order to increase economic attractiveness, a cluster involving several mills can be developed (Marinova et al., 2010b). A pulp mill or a chemical plant will be used as the center of the cluster where hemicellulose prehydrolysate will be collected from several mills and converted to other products (furfural and ethanol). In the case under study, the dissolving pulp mill is considered as the center of a cluster, where 7,000 t/d of hemicellulose

Table 3.6 Main Applications of Xylan	
Pulp and paper	Beater additive
	Improved swelling
	Porosity
	Drainage
	Strength
	Fiber coating
	Wood resin stabilizer
Food	Xylose
	Xylitol
	Biodegradable polymers
	Plastics, films
	Coatings with increased hydrophobicity and water resistance (acetyl xylans)
Fermentation	Enzymes
	Xylanase
	Xylose isomerase
	Biopolymers
	Polyhydroxy alkanoates
Chemical	Thermoplastic material
	Polypropylene filler
	Paint formulations
	Gel-forming material
	Chiral polymer building blocks
Pharmaceutical	Anticoagulant
	Anticancer agent
	Cholesterol reducing agent
	Wound treatment agent
	HIV inhibitor
Source: *Based on Christopher (2012)*.	

prehydrolysate are collected and converted to ethanol. The steam consumption of the dissolving pulp mill and the ethanol plant before energy optimization are 30.24 and 5.184 GJ/adt, respectively (energy values based on the production of the dissolving pulp mill). The application of pinch analysis results in maximum steam savings of 30% of the current consumption for the dissolving pulp mill and 43% for the ethanol plant. An analysis of the thermal profiles (grand composite curve) of the integrated site confirms it is possible to recover 2.11 GJ/adt of waste heat from the ethanol plant (condensers of the distillation towers) that can be transported to the dissolving pulp mill. This is possible because the pinch point temperature (the lowest temperature difference between the curves representing the heat demands and sources of the process) of the ethanol plant (89°C) is higher than in the dissolving pulp mill (53.7°C). The minimum energy demand of the integrated site is 22 GJ/adt. As the steam production capacity of the mill has been reduced from 30.24 GJ/adt to 27.57 GJ/adt, there is still 5.57 GJ/adt of excess steam. However, the application of the complete unified methodology is required to ensure that the maximum steam savings are achieved.

Preextraction of hemicellulose can provide a totally new feedstock for biofuel/bioethanol production, thus increasing the total revenue stream for the pulp and paper industry (van Heiningen, 2006; Ragauskas et al., 2006). It is therefore desirable to develop a pretreatment process that can solubilize hemicellulose sugars with minimal formation of fermentation inhibitors, while preserving the fiber integrity.

It is expected that preextraction of these "waste" hemicelluloses prior to kraft pulping could substantially improve pulp mill operations (Ragauskas et al., 2006; Thorp and Raymond, 2005). The benefits of hemicellulose preextraction are:

- Reduction in kraft cooking times.
- Enhancing kraft cooking liquor impregnation.
- Improved pulp properties.
- Improving pulp production capacity for kraft pulp mills that are recovery-furnace limited.

These process benefits and production of biofuels are strong drivers for the development of wood hemicellulose preextraction technologies for kraft pulp mills (Bajpai, 2012). Any preextraction of wood chips prior to kraft pulping needs to develop a system that is readily integrated with modern pulping operations and will not deteriorate the quality of kraft pulps. A key physical parameter in the production of many grades of paper is the strength of the final paper sheet. It has been well documented that if the Degree of polymerization (DP) of cellulose is decreased beyond its normal $\sim 1,600$ post-pulping to ~ 700 after bleaching (Yanagisawa et al., 2005), the strength properties of the sheet are degraded. This relationship is due to the fact that cellulose is the primary load-bearing element in a lignocellulosic fiber and has a direct relationship to the fiber strength, which contributes to paper strength. Hence, any hemicellulose preextraction technology employed prior to kraft pulping needs to minimize the hydrolysis of cellulose. Furthermore, it has been reported that hemicellulose content is related to paper bond strength, which has been attributed to the adhesive properties of hemicellulose. Studies suggest that for kraft pulps with an α-cellulose content higher than ∼80%, a decrease in paper sheet strength properties occurs (Molin and Teder, 2002; Page and Seth, 1985; Schönberg et al., 2001). This product specification defines a limit for hemicellulose preextraction technologies.

In an ideal situation, if one could extract 15–20% hemicellulose before pulping and get the same pulp yield as obtained before, it will be possible to keep the same pulp mill production level without increasing the wood demand and would also reduce black liquor solids (BLS) going to the recovery boiler. Removing the recovery boiler bottleneck may allow the manufacturing of more tonnage, which will further improve the profitability of the kraft mill.

Pretreatment with dilute sulfuric acid with pH control using ammonia and or lime is found to be a good approach for hemicelluloses preextraction. In this method, higher hemicellulose yields up to 90% are obtained and the cost is lower (Huang et al., 2008). Dilute acid (0.5–1.0% sulfuric) at moderate temperatures (140–190°C) can recover most of the hemicelluloses as dissolved sugars (Knappert et al., 1981). Water extraction at higher temperatures (200–230°C) can completely recover hemicellulose from hardwoods and herbaceous materials, without significant degradation (Mok and Antal 1992).

Saska and Ozer (1995) found that hemicellulose from sugarcane bagasse can be efficiently extracted with water as the extractant. About 89% the original amount of xylose was recovered under the operating conditions of the solid/liquid ratio at 1:5, the extraction temperature at 150–170°C, and the extraction time of 15–30 min. The main advantages of the water extraction method over the dilute acid pretreatment are:

• lower corrosion to equipment
• less xylose degradation
• less byproducts including inhibitory compounds in the extracts
• more easier recovery of acid from the hydrolyzate.

When particle size reduction was used prior to extraction, the aqueous process gave almost 90% recovery of xylose, superior to steam explosion-based extraction (Saska and Ozer, 1995).

The use of water as prehydrolysis stage relies on the *in situ* hydrolysis of acetate groups on the hemicellulose chains yielding acetic acid.

The liberated acid lowers the solution pH to a range of 3–4. This results in the hydrolysis and solubilization of hemicelluloses. Control of the prehydrolysis parameters is an important consideration, as more vigorous conditions will degrade the fiber resource.

Water prehydrolysis is found to be more effective at removing hemicelluloses than steam prehydrolysis, especially for softwoods. All prehydrolysis treatments also extract low levels of lignin and extractives. A key consideration for extracting hemicelluloses prior to kraft pulping for nondissolving grades of paper is the need to yield a wood furnish that still yields excellent physical strength pulp properties. This will undoubtedly require an optimization of hemicellulose preextraction technologies providing optimal removal of hemicelluloses for biofuel production and sufficient retention of select hemicelluloses for the production of high quality kraft pulps.

Microwave heat-fractionation of wood has also been studied to extract hemicelluloses (Lundqvist et al., 2002; Palm and Zacchi, 2003). This method requires a treatment temperature of 180–200°C for 2–5 min.

Steam explosion is reported to be an effective pretreatment for hemicellulose hydrolysis (Cara et al., 2006; Ibrahim and Glasser, 1999; Josefsson et al., 2002; Shimizu et al., 1998). In steam explosion process, biomass is pretreated by pressurized steam followed by rapid relieving of pressure. This breaks down the lignocellulosic structure so that the lignin is readily depolymerized. As a result of this, the hemicellulose is easily hydrolyzed (Cara et al., 2006). The steam explosion process can result in around 50% insoluble residue of the wood, consisting mainly of cellulose. The remaining portion which chiefly contains hemicelluloses and lignin, can be recovered with alkali extraction (Josefsson et al., 2002).

Shimizu et al. (1998) conducted steam explosion of different species of hardwood chips at 180–308°C for 1–20 min. This resulted in partial hydrolysis of hemicelluloses. The resulting sugars were extracted with water. The xylose yield was 10–20% of the starting materials.

Ibrahim and Glasser (1999) used steam treatment to break down and separate the red oak wood chips into fibers and polymer products. This resulted in nearly complete recovery of xylan and obtaining almost hemicellulose-free pulps.

Josefsson et al. (2002) used the steam treatment to fractionate the aspen wood components, for obtaining a high cellulose yield and an appropriate molecular weight distribution, while recovering hemicelluloses. Pulps with different xylan contents ranging from less than 1% to 7% and different molecular weight of cellulose ranging from less than 40,000 to 900,000 were prepared at varying time and temperature conditions.

The steam explosion method is found to be more environmental friendly in comparison with alternative methods. It requires lower capital investment (Cara et al., 2006). However, this method has a disadvantage in that it is difficult to restrain fibers from fragmentation.

Tucker et al. (2003) studied the combined dilute acid−steam explosion method for biomass treatment. Corn stover was subjected to 1 wt% H_2SO_4 for 70−840 s in a steam explosion reactor at 160°C, 180°C, and 190°C. The obtained yields of xylose were 63−77% at 160−180°C, and more than 90% at 190°C.

Modified twin-screw extruder was used by N'Diaye et al. (1996) and N'Diaye and Rigal (2000) to extract hemicelluloses from *Populus tremuloides* with a 5% NaOH solution as extracting solvent. This extruder, called a thermomechanicochemical fractionation system, allows the integration of extrusion, cooking, liquid−solid extraction, and liquid/solid separation (filtration) in a single step, and operates in a continuous mode. With such a reactor, alkaline extraction can be operated at a lower L/S ratio (six times less than a batch reactor) and lower residence time, and 90% of the initial hemicelluloses can be recovered.

Gabrielii et al. (2000) extracted hemicelluloses from aspen wood (*Populus tremula*) by alkali extraction combined with hydrogen peroxide treatment, ultrafiltration, and recovery by spray drying. The aspen wood was first cut and refined. The resulting fiber suspension was treated with a dilute HCl solution to swell the fibers. The fibers were then cooled and ammonium hydroxide was added to dissolve the pectins, stirred overnight, and centrifuged to remove pectins, starch, and fat. The residue was subjected to a NaOH (1%)−ethanol (70%) solution in order to solubilize lignin. After centrifugation, hemicellulose was extracted with 4% NaOH. The combined two filtrates were bleached with peroxide to minimize the residual lignin content. The suspension was finally ultrafiltered and spray-dried to obtain hemicellulose.

Sun et al. (2001) investigated the extraction of hemicelluloses from fast-growing poplar wood. In their process, poplar wood chips were dried and dewaxed by extraction with toluene–ethanol mixture (2:1, v/v). This was followed by partial delignification with an acidic NaCl solution. Hemicelluloses were extracted with 1.5–8.5% sodium hydroxide leading to 65.6–89.3% solubilization of the original hemicelluloses. The slurry was filtered into solid (cellulose) and filtrate (hydrolyzates). The hydrolyzates were neutralized to pH 5.5, and the hemicelluloses were precipitated in 95% ethanol and filtered, washed, and dried. Alkali extraction with sodium hydroxide was used in all these three processes, and this method was found to be effective in extraction of hemicelluloses from hardwood. The hemicelluloses concentration of filtered hydrolyzates was found to be very low, e.g., 2–3% depending on the solid/solvent ratio used. So, it is very important to concentrate hemicelluloses in order for efficient subsequent fermentation into ethanol or xylitol, or to obtain pure hemicelluloses for other uses, for example, in making new biobased materials. Precipitation with ethanol was used for separation of hemicelluloses in the processes used by N'Diaye et al., (1996) and Sun et al. (2001), while ultrafiltration was used in the process used by Gabrielii et al. (2000). Methods used by N'Diaye et al. (1996) and Sun et al. (2001) involved acidification or neutralization with acid to adjust pH to 5–5.5 suitable for precipitation of hemicelluloses with ethanol. In this process, ethanol and acid are additional chemicals required and additional equipments for recovery of ethanol is necessary, thus leading to increase in production and capital cost. In addition, some ethanol is probably included in the precipitated solid phase representing ethanol loss or additional separation cost. In the latter process (Gabrielii et al., 2000) which uses ultrafiltration, the operating cost and capital cost are lower. As to reactors, a twin-screw extruder representing a highly efficient continuous operation is used in the process used by N'Diaye et al. (1996). Basically, the water extraction method, which represents a mild pretreatment, brings about a higher molecular weight hemicellulose (Ragauskas, 2006). Twin-screw extruder combined with ultrafiltration may be a good choice for isolation of hemicelluloses by water extraction.

Schlesinger et al. (2006) found nanofiltration to be much better than ultrafiltration for separating hemicelluloses from hydrolyzates by alkaline method. They studied the performance of four polymeric nanofiltration and one tight ultrafiltration membranes for isolating hemicelluloses from alkaline process liquors containing 200 g/l NaOH. The experimental results showed that the hemicelluloses of molar mass over 1,000 g/mol are almost retained. In addition, two of the membranes with the nominal molecular weight cutoff (MWCO) of 200–300 and 200–250 g/mol, respectively, are most efficient in retention of up to 90% of hemicelluloses, while the tight ultrafiltration membrane with MWCO of 2,000 g/mol exhibited less than 70% retention of hemicelluloses. Ali et al. (2005) patented an alkaline treatment system for recovering hemicelluloses where prefiltration units with a screen size of 400–650 mesh, followed by one nanofiltration membrane, was able to retain compounds with a molecular weight of about 200 and higher. Therefore, nanofiltration is an excellent separation procedure for recovery of hemicelluloses from hydrolyzates, and the combination of a twin-screw extruder and nanofiltration can be considered to be the best selection for extracting hemicelluloses from hardwood chips.

van Heiningen (2005) preextracted hemicellulose from mixed hardwood chips with pure water and 10% alkaline solutions. The extracted chips were then subjected to kraft cooking at standard conditions. Results showed that with these two approaches, the pulp yield decreased 5–7%, although 10% organics could be extracted. To avoid this problem, they employed a new method that can increase the pulp yield to the same level or even 1% higher than that of a kraft cook control (no extraction). This method was shown to result in other benefits:

1. A 3% reduction of effective alkali (EA) charge in the digester.
2. A 40% increase in delignification rate.
3. Rejects reduction at higher kappa numbers.
4. An 8% decrease in organic load to the recovery boiler, based on o.d. (oven dried) wood.

Organosolv fractionation technology developed by National Renewable Energy Laboratory utilizes a ternary mixture of methyl isobutyl ketone, ethanol, and water in the presence of low concentrations of sulfuric acid to effect a separation of cellulose, hemicellulose, and lignin. The method typically requires a treatment temperature of

140°C for 1 h. This approach has worked well to fractionate hard-woods, yielding high purity cellulose and selectively dissolving lignin and hemicellulose (Bozell et al., 1995). However, the method proves difficult with softwoods, requiring more acid, higher temperatures, and longer retention times, resulting in poor cellulose pulps. For integration into a kraft biorefinery, the organosolv extraction method would need to be studied further.

3.2 PREEXTRACTION OF ANTIOXIDANTS

Naturally occurring antioxidants (phenolics or polyphenolics) can also be produced from lignocellulosic wood materials in addition to hemicelluloses. Antioxidants could be used as a cheap, renewable food additive (Cruz et al., 1999; Gonzalez et al., 2004). Gonzalez et al. (2004) subjected *Eucalyptus globules* wood chips to acid hydrolysis with 2.5–5% H_2SO_4 at a liquid/solid ratio of 8:1 g/g and 100–130°C. The resulting slurry was vacuum filtered into hydrolyzates and a solid consisting of cellulose and lignin. Then antioxidants were extracted from hydrolyzate with ethyl acetate as solvent. The resulting organic phase was vacuum evaporated to remove and recycle ethyl acetate to the extractor, leaving the antioxidants-containing extracts behind. The aqueous phase from extraction contains xylose (representing hemicellulose) which can be fermented to yield ethanol and/or xylitol with different yeasts.

3.3 BLACK LIQUOR GASIFICATION

Approximately half of the wood, which is used as feedstock in the pulp and paper industry, is converted into a highly viscous liquor called black liquor in the pulping process. It is traditionally burned to recover cooking chemicals and produce process steam and power for the mill (Bajpai, 2008; Grace, 1987, 1992; Grace and Timmer, 1995; Hupa et al., 1994). Black liquor has properties uniquely suitable for gasification:

- It is liquid and easily pumped into the pressurized gasifier.
- The liquid form makes it easy to atomize into fine droplets.
- It is highly reactive due to high sodium and potassium content.

These properties make the gasification of black liquor easier and more rapid than for any other biomass feedstock. Black liquor gasification (BLG) should be an integral part of an Integrated forest biorefinery (IFBR) because its process heat may be used in the sugar conversion unit operations, and the synthesis gas may be used to replace fossil fuels, in particular oil in the lime kiln. The gasification synthesis gas may be used as feedstock to produce transportation fluids such as Fisher–Tropsch liquid hydrocarbons, methanol, and mixtures of higher alcohols. The key requirement for implementation of BLG is to demonstrate the reliability and efficiency of the technology at commercial scale while regenerating the pulping chemicals.

The developments in gasification technology have been conducted over years for efficient recovery of biobased residues in the chemical pulp mills (Ådahl et al., 2004; Dahlquist et al., 2009; Ekbom et al., 2003; Farmer and Sinquefield, 2003; Harvey and Facchini, 2004; Larsen et al., 1998, 2000, 2003; Larsen et al., 2006a; Möllersten et al., 2003a,b; Sricharoenchaikul, 2001; Yan et al., 2007). BLG has been a popular topic in several conferences on biorefining, engineering, pulping, and environmental matters. Several studies have been conducted to analyze the technical, economic, and climate change mitigation performance of gasification process and a number of pilot plants have been successfully operated. There has been more focus on possible integration of gasification process for increased energy self-sufficiency. This serves as a base for a modern biorefinery concept at the pulp mills, coproducing pulp, and valuable energy products. To add more value to the integrated pulp mills, the research and development of BLG has been focused on: increased power production from integrated BLG system switching a pulp mill from a net electricity importer to exporter; utilization of surplus BL energy for biofuel production and potentially converting a modern kraft mill to become biofuel supplier in the future energy system; improved performance of the combined heat and power systems using integration of the BLG with a gas turbine; evaluation of advanced power cycles for potential increase in electricity surplus, including combined cycle (CC), steam injected gas turbine (STIG) cycle, evaporative gas turbine or humid air turbine (EvGT or HAT) cycles; assessment of BLG technology in terms of technical, economic, and climate change mitigation; integration of pulp mills with CO_2 mitigation technologies; cost-competitiveness of electricity or biofuel production via BLG; and resolving the material challenges of BL gasifier refractory lining.

The two main technologies under development are pressurized and atmospheric gasification, being commercialized by Chemrec AB and ThermoChem Recovery International (TRI), respectively.

BLG offers a way to generate electricity and to reclaim pulping chemicals from black liquor. This is accomplished by converting the fixed carbon to a combustible gas mixture using oxygen-containing gases such as oxygen, carbon dioxide, and water vapor. The combustible gas is then burned to generate electrical power. BLG would replace the Tomlinson recovery boiler for the recovery of spent chemicals and energy. Gasification may become part of integrated gasification and combined cycle (IGCC) operation, or lead to pulp mills becoming biorefineries (Larsen et al., 2003).

The organic matter in black liquor is partially oxidized with an oxidizing agent to form syngas in the gasifier, while leaving behind a condensed phase. The syngas is cleaned to remove particulates and tars and to absorb inorganic species (i.e., alkali vapor species, SO_2, and H_2S), and this is done to prevent damage to the gas turbine and to reduce pollutant emissions. The clean syngas is burned in gas turbines coupled with generators to produce electricity, and gas turbines are inherently more efficient than the steam turbines of recovery boilers due to their high overall air fuel ratios (Nilsson et al., 1995). The hot exhaust gas is then passed through a heat exchanger (typically a waste-heat boiler) to produce high-pressure steam for a steam turbine and/or process steam. The condensed phase (smelt) continuously leaves the bottom of the gasifier and must be processed further in the lime cycle to recover pulping chemicals.

Essentially, all of the alkali species and sulfur species leave in the smelt (mostly as Na_2S and Na_2CO_3) in the recovery boilers, but in gasifiers, there is a natural partitioning of sulfur to the gas phase (primarily H_2S) and alkali species to the condensed phase after the black liquor is gasified. Because of this inherent separation, it is possible to implement alternative pulping chemistries that would yield higher amounts of pulp per unit of wood consumed (Larsen et al., 1998, 2003). Gasification at low temperatures thermodynamically favors a higher sodium/sulfur split than gasification at high temperatures, which results in higher amounts of sulfur gases at low temperatures. Because a large amount of the black liquor sulfur species leaves the low-temperature process as H_2S, H_2S may be recovered via absorption to facilitate alternative pulping chemistries. Industry has numerous patented processes for accomplishing the absorption, including using green or white liquor as an absorbing solvent (Larsen et al., 1998, 2003; Martin et al., 2000).

The partitioning of sodium and sulfur in BLG requires a higher capacity for the lime cycle compared with the current technology. The sodium/sulfur split results in a higher amount of Na_2CO_3 in the green liquor because less sulfur is available in the smelt to form Na_2S. For each mole of sulfur that goes into the gas phase, one more mole of Na_2CO_3 is formed in the condensed phase (Larsen et al., 2003). The increase in Na_2CO_3 results in higher causticization loads, increases in lime kiln capacity, and increases in fossil fuel consumption to run the lime kiln. This leads to higher raw material and operating costs, which must be reduced in order to make the gasification process economically favorable.

BLG may be performed either at low temperatures or at high temperatures, based on whether the process is conducted above or below the melting temperature range (650–800°C) of the spent pulping chemicals (Sricharoenchaikul, 2001). In low-temperature gasification, the alkali salts in the condensed phase remain as solid products, while molten salts are produced in high-temperature gasification. Low-temperature gasification is advantageous over high-temperature gasification because gasification at low temperatures yields improved sodium and sulfur separation. Additionally, low-temperature gasification requires fewer constraints for materials of construction because of the solid product. However, the syngas of low-temperature gasification may contain larger amounts of tars, which can contaminate gas cleanup operations in addition to contaminating gas turbines upstream of the gasifier. These contamination problems can result in a loss of fuel product from the gasifier (Sricharoenchaikul, 2001).

3.3.1 BLG Technologies
BLG technologies can be divided into two major classes:

1. *Low-temperature gasification*: Low-temperature gasifier operates at 600–850°C, below the melting point of inorganics, thus avoiding smelt–water explosions (Patrick and Siedel, 2003).
2. *High-temperature gasification*: High-temperature gasification units generally operate in the range of 900–1,000°C and produce a molten smelt (Patrick and Siedel, 2003).

Several companies have performed trials to develop a commercially feasible process for BLG. The history of BLG development is well described by Whitty and Baxter (2001) and Whitty and Verrill (2004). Only two technologies are currently being commercially pursued: the Manufacturing and Technology Conversion International (MTCI) (low temperature) and Chemrec (high temperature) technologies. Weyerhaeuser, New Bern, uses a Chemrec booster for BLG, but it operates at atmospheric pressure, which does not give maximum energy efficiency. Energy efficiency is enhanced by going to higher pressures. Trials were run at Kappa Kraftliner, Sweden, in which the black liquor was gasified at high temperature and pressure in a reactor then the gas was cooled and separated from droplets of smelt. The condensate was dissolved to form low-sulfidity green liquor. The raw gas containing carbon monoxide and carbon dioxide was saturated with steam at high pressure then cooled and stripped of particles. The gas can be used as a feedstock in a CC technology or for chemical synthesis (Larson et al., 2000).

3.3.1.1 SCA-Billerud Process
This process was developed by SCA Billerud and was implemented at Ortviken (Dahlquist and Jacobs, 1994). It is a low-temperature pyrolysis process recovering calcium−sulfite black liquor in 1958. The plant was in operation for 16 years as the recovery system for a calcium−sulfite system. The process operates at 725°C, and the synthesis gas is quenched in a heat exchanger to 150°C (Whitty, 2005). The cooled synthesis gas is scrubbed and cleaned in cyclone separators to remove solid particulates. The residence time is too short, resulting in a significant amount of residual carbon in the dust. The process operates with high availability and less manpower requirement when compared to the recovery boiler. The disadvantage of the process is low carbon conversion due to very short residence time in the reactor, and also low thermal energy efficiency (Whitty, 2005; Whitty and Verill, 2004). The technology has been abandoned due to technical inferiority. It was operated until 1980.

3.3.1.2 Direct Alkali Regeneration System Process
This process performs direct causticization of black liquor and shows promise of providing a simpler black liquor recovery technology that is suitable for small scale pulp mills (Kulkarni et al., 2007). The direct causticization leads to conversion of sodium carbonate to sodium

hydroxide during BLG process when smelt from gasifier is dissolved in water. This eliminates the requirement of separate causticizing unit with an energy-intensive lime kiln. The first commercial direct alkali regeneration system process plant operated at the APPM mill in Burnie, Tasmania (Grace, 1987). The recovery system processed 320 t/d of BLS, producing 95 t/d of sodium hydroxide. The sodium carbonate reacts with particles of ferric oxide in a fluidized bed combustor to produce sodium ferrite and subsequently to obtain sodium hydroxide by hydrolyzing the sodium ferrite. This process operates autogenously (with no auxiliary fuel) when firing liquors as low as 40%. Unlike the smelt in conventional recovery cycle, the combustion product is solid, so the process is safe (Kulkarni et al., 2007). However, one drawback to the process is to control the oxidation level of iron oxide, so that no inert compounds are formed. Another problem is the formation of large amounts of iron oxide dust which proved unsuitable for commercial pulp mills.

3.3.1.3 BLG with Direct Causticization

ABB developed a new BLG technology with direct causticization for energy optimization during the 1990s (Dahlquist and Jacobs, 1994). In this process, a circulating fluidized bed of titanium dioxide is injected with black liquor under pressurized conditions (5 bar) and temperature range 650–720°C. The titanium dioxide reacts with sodium carbonate to form sodium titanate and carbonate is converted to carbon dioxide that results in direct causticization. First, sodium sulfate is reduced to sodium sulfide, and then sodium sulfide is stripped off as hydrogen sulfide. The hydrogen sulfide is absorbed from the raw gas in a selective absorber, using part of the white liquor for absorption. Some interesting studies on BLG process with direct causticization have been presented by Dahlquist and Jacobs (1994), Dahlquist (2003), Nohlgren (2004), and Nohlgren and Sinquefield (2004), Zou et al. (1992), Zeng and van Heiningen (1996).

Most of the studies showed that the direct causticization has several advantages over the conventional black liquor recovery system including lower capital cost for fewer process steps, e.g., no requirement for the lime kiln (Sinquefield, 2005; Warnqvist et al., 2001). A higher concentration of white liquor can be achieved, and the production of white liquor containing different sulfur contents. In terms of energy efficiency, the direct causticization process is better than conventional recovery system; as only 25% of the energy demand in conventional recovery is required for direct causticization (Richards et al., 2002).

3.3.1.4 MTCI Fluidized Bed Gasification

MTCI has developed a pulse-enhanced, indirectly heated fluidized bed gasifier that is used at Norampac and Georgia Pacific pulp mills (Patrick and Siedel, 2003). MTCI technology is also known as TRI (Durai-Swamy et al., 1991; Mansour et al., 1992, 1993, 1997; Rockvam, 2001; Whitty and Verrill, 2004). MTCI projects are running in mills with a Na_2CO_3 semichemical cooking process. It operates at $600-620°C$. This technology has some significant advantages (Whitty, 2005):

- efficient heat transfer with uniform heat flux
- high combustion efficiency
- low NO_x emissions
- absence of moving parts
- pressure boost help exhaust gases to flow through a superheater and heat recovery steam generator (HRSG).

In the MTCI process, the bed is indirectly heated by several bundles of pulsed combustion tubes, which burn some of the produced gas. Black liquor is sprayed into the fluidized bed and coats the solids, where it is quickly dried and pyrolyzed. The remaining char reacts with steam to produce a hydrogen-rich fuel gas (Rockvam, 2001). Part of the bed material is continuously removed, dissolved in water, and cleaned from unburned carbon to obtain green liquor. The produced gas is passed through a cyclone to separate solids and then to an HRSG. Part of the generated steam is used in the gasifier as both reactant and fluidizing medium. The gas continues through a Venturi, a gas cooler, and is finally cleaned from H_2S in a scrubber with some of the green liquor. The cleaned gas contains about 73% H_2, 14% CO_2, 5% CH_4, and 5% CO (Rockvam, 2001). The heating value of the gas is high (~ 13 MJ/Nm^3). It can be burned in an auxiliary boiler, used in a fuel cell to generate electricity and when pressurized it can be fired in a gas turbine.

There are two MTCI projects running. The first one is running at Georgia Pacific Corporation's Big Island mill in Virginia, USA. This system is a full-scale gasifier, designed to process 200 tds/d and is fully integrated with the mill (DeCarrera, 2006). The second project is running at Norampac Trenton mill, Ontario, Canada (Middleton, 2006;

Newport et al., 2004; Vakkilainen et al., 2008). Prior to the start-up of the gasifier, the mill had no chemical recovery system. For over 40 years, the mill's spent liquor was sold to local counties for use as a binder and dust suppressant on gravel roads. The discontinuance of the spreading of spent liquor required Norampac to select, purchase, and install a technology to process spent liquor. The TRI BLG system was selected. The capacity of the TRI spent liquor gasification system is 125 t/d of BLS. TRI's scope of supply included the steam reformer, pulse combustors and fuel train, detailed engineering and start-up support, materials handling equipment, and instrumentation. The project, which started operations in 2003, is operating day in and day out, meeting all of the needs of the mill's chemical recovery requirements. Process optimization is continuing in the area of energy recovery. TRI's gasification process is ideal for use in a forest products biorefinery as it is uniquely configured for high-performance integration with pulp and paper facilities and is capable of handling a wide variety of cellulosic feedstocks, including woodchips, forest residuals, agricultural wastes, and energy crops, as well as mill byproducts (spent liquor). Compared to other biomass gasification technologies that are based on partial oxidation, TRI's steam reformer converts biomass to syngas more efficiently, producing more syngas per ton of biomass with a higher Btu content. This medium Btu syngas can be used as a substitute for natural gas and fuel oil, and as a feedstock for the production of value-added products such as biodiesel, ethanol, methanol, acetic acid, and other biochemicals. TRI's technology can be integrated with a wide variety of catalytic and fermentation technologies to convert the syngas to high-value biobased fuels and chemicals. For example, syngas generated by TRI's technology can be conditioned and sent to a commercially proven gas-to-liquids (GTL) facility (i.e., Fischer–Tropsch or other catalytic technologies) inside the biorefinery. The GTL process produces a range of products (naphtha, gasoline, diesel/kerosene, wax, methanol, and dimethylether (DME)) that are stabilized for storage and transported offsite to a downstream refinery for conversion to marketable products. The unreacted syngas and light noncondensable gases (tail gas) are utilized in the process to replace fossil fuels. Additionally, the GTL conversion, which is exothermic, provides another source of process heat that is recovered and used. A fully integrated forest products biorefinery utilizing TRI's technology will achieve thermal efficiencies from 70% to 80% depending upon process configuration and biomass feedstock. Figure 3.1 shows schematic of MTCI steam reformer.

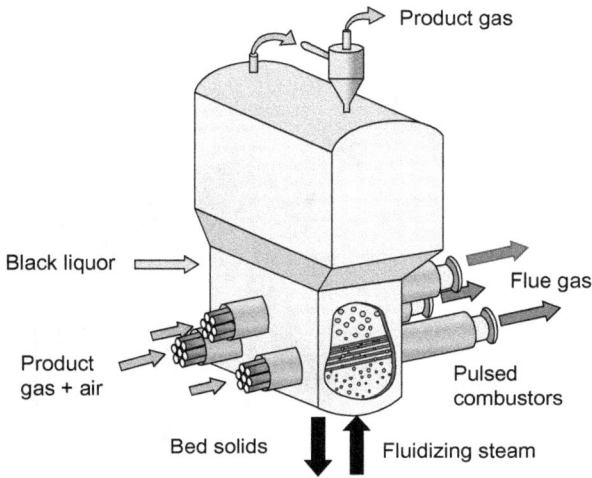

Figure 3.1 Schematic of MTCI steam reformer. Based on Whitty and Baxter (2001).

The MTCI system is also in the trial stages at Georgia Pacific, Big Island. Technical issues have included excessive tar formation (over 30% of the organic content of the processed liquor was lost to the sewer as tar), lower than expected carbon conversion (approximately 80% vs the expected 99%) and concerns about the design of the fluidization system.

A major drawback with the MTCI process is low carbon conversion due to low temperature, but still the thermal efficiency is well above 70% as compared to 65% or less for conventional recovery boilers (Suresh, 2002).

3.3.1.5 Chemrec Gasification

Currently, the most commercially advanced BLG technology is the Chemrec technology, which is based on entrained-flow gasification of the black liquor at temperatures above the melting point of the inorganic chemicals. Chemrec is working on both an atmospheric version and a pressurized version of a high temperature downflow entrained-flow reactor. The atmospheric versions are mainly considered as a booster to give additional black liquor processing capacity. The pressurized version is more advanced and would replace a recovery boiler or function as a booster (Brown and Landälv, 2001; Kignell, 1989; Stigsson, 1998; Whitty and Nilsson, 2001; Whitty and Verrill, 2004).

Operates as a booster unit to
the recovery boiler in the
Weyerhaeuser pulp mill in
New Bern, North Carolina

Corresponds to ~15% of the
mill's production

Start up	:	Dec 1996
Capacity	:	330 tDS/day
		50 MW
Oxidant	:	Air
Pressure	:	Atmospheric
Operation	:	> 40,000 h

Figure 3.2 Chemrec's atmospheric booster in New Bern, NC, USA. Reproduced with permission from Landälv (2007).

3.3.1.5.1 Atmospheric System

In this system, black liquor is fed as droplets through a burner at the top of the reactor. The droplets are partially combusted with air or oxygen at $950-1,000°C$ and atmospheric pressure. The heat generated sustains the gasification reactions. The salt smelt is separated from the gas, falls into a sump, and dissolves to form green liquor. The produced gas passes a cooling and scrubbing system to condense water vapor and remove H_2S. The gas has low heating value (~ 2.8 MJ/Nm^3) and is suitable for firing in an auxiliary boiler. It consists of $15-17\%$ CO_2, $10-15\%$ H_2, $8-12\%$ CO, $0.2-1\%$ CH_4, and $55-65\%$ N_2 (Lindblom, 2003). The thermal efficiency is quite low. An atmospheric Chemrec booster system with a firing rate of 270 tds/d has been in use at Weyerhaeuser's New Bern mill since 1997 (Fig. 3.2). It was shut down in 2001 due to extensive cracking in the reactor shell, and it was started again in 2003. The gasifier was rebuilt with a new reactor vessel as well as a modified refractory lining design and it has operated well since then (Brown et al., 2004).

3.3.1.5.2 Pressurized System

The pressurized Chemrec BLG system is a new technology for energy and chemicals recovery in pulping processes with the aim of offering pulp mills significant cash flow additions through increased utilization of the energy content in its renewable feedstocks for production of green electricity, automotive fuels, or hydrogen. In addition, the Chemrec system offers the option of utilizing advanced new cooking processes resulting in increased pulp yield and giving a fundamentally different opportunity to manage the environmental impact of black liquor conversion. The capital outlay for the system is larger than for the corresponding recovery boiler, but the extra investment is rapidly repaid through the increased cash flow that is generated.

The core of the pressurized Chemrec system is the gasifier unit, an entrained-flow reactor, where concentrated black liquor is gasified at approximately 32 bar pressure and at a temperature of about $1,000°C$. Oxygen is used as the oxidant. In the Chemrec DP-1 plant, operations start with a refractory lined reactor, with the option of later switching to a reactor with a cooling screen making up the containment. Black liquor reacts in the reaction zone to form smelt droplets consisting of sodium and sulfur compounds and a combustible gas mainly consisting of CO, H_2, and CO_2. Part of the sulfur in the black liquor also ends up as H_2S in the gas. The smelt droplets and the combustible gas are fed to a quench vessel where they are cooled when brought into direct contact with condensate, which is recycled from the downstream gas cooler. The smelt droplets are separated from the gas and dissolved in the quench liquid to form a green liquor solution. The green liquor, after heat exchanging with weak wash and final cooling, is fed to battery limit. The gas leaving the primary quench is further cooled to saturation in a second quench device. The saturated gas is simultaneously scrubbed from particles and cooled in a counter current condenser (CCC). The DP-1 plant utilizes cooling water as cooling medium in the CCC. In mill applications, the CCC will instead produce medium-pressure and low-pressure steam. The cooled gas is cleaned from the sulfur in multistage short time contactors that utilize white liquor as the absorbing liquid, after which the purified and cooled syngas is burned in a flare. The Chemrec reactor is designed to achieve high carbon conversion and sulfur reduction, exceeding what is normally obtained in a recovery boiler. The quantity of unburned carbon and sulfate in the green liquor is consequently low. The Chemrec DP-1 plant is operated from its own control room through a state-of-the-art computer-based steering and control system. Media such as black liquor, white liquor, steam, water, and electricity are supplied from the

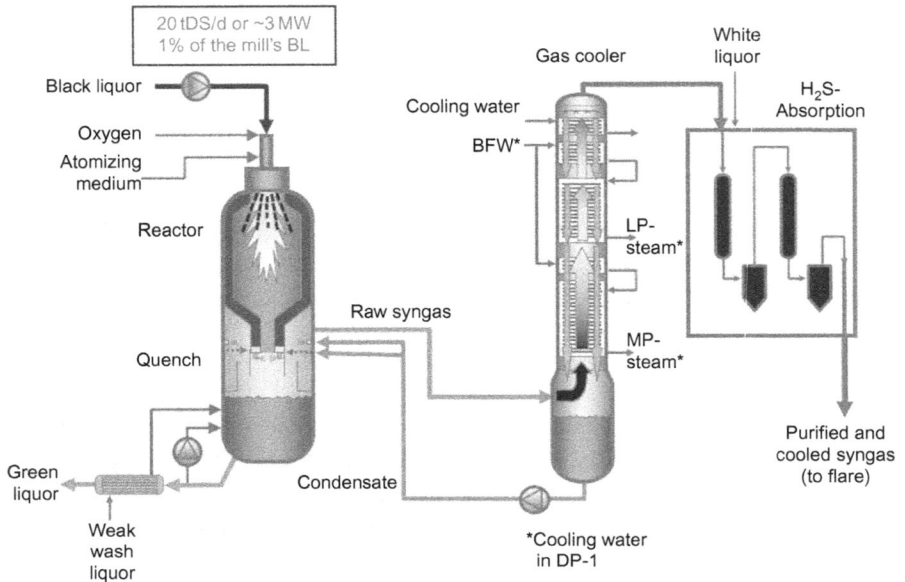

Figure 3.3 High temperature O$_2$-blown plant Pitea, Sweden. Reproduced with permission from Landälv (2007).

adjacent Kappa Kraftliner mill and green liquor is delivered back to the mill. The DP-1 plant operation is managed by Chemrec and during longer periods of operation supported by staff from the Kappa Kraftliner and SCA Packaging Munksund mills (Chemrec, 2009).

A major development in Chemrec gasification is the installation of an oxygen-blown, pressurized black liquor gasifier in Piteå, Sweden (Figs 3.3 and 3.4). Chemrec has operated the pressurized O$_2$-blown plant DP-1 in Piteå for some 12,000 h (on its own) and the combined gasifier and the BioDME plant for another 6,000 h (Landälv, 2012). The synthesis gas will be used to produce second-generation green automotive fuels. The pilot plant handles 20 Mt BLS/day on a fully continuous basis and has an operating pressure of 30 bar. The results from the pilot plant are being used in development of a full commercial size gasifier for 500 Mt of BLS/day. The facility provides high industrial standards and is equipped with modern process and data control system. The produced gas has been determined to contain about 41% H$_2$, 31% CO$_2$, 25% CO, 2% CH$_4$, and 1.4% H$_2$S (Lindblom, 2006). Some interesting studies in Chemrec BLG system are discussed below.

Figure 3.4 DP-1 plant location in Pitea, Sweden. Reproduced with permission from Landälv (2007).

Larsson et al. (2006a,b) determined the inaccuracies in thermochemical data that influenced process variables resulting from equilibrium modeling of Chemrec oxygen-blown pressurized gasification process. The effect of the variation in pressure in the gasifier (25–32 bar) had a small effect on H_2S formation, but the data uncertainties became larger in temperature variations higher than 1,000°C. A higher moisture content in both recovery system and gasifier favored NaOH(g) and KOH(g) formation. The impacts on sulfur equilibrium chemistry were found to be of utmost importance. The calculation results indicated significant uncertainties in Na_2S(s,l), K_2S(s,l), and K_2CO_3(s,l) for KOH(g) formation. Marklund et al. (2007) identified the four most important parameters for a proposed computational fluid dynamics (CFD) modeling of Chemrec BLG as an initial step before a complete model validation against experimental data is done. The considered performance response parameters were: (1) fraction of volatile sulfur, (2) sulfide to sulfate ratio, (3) fraction of volatile carbon, and (4) CO to CO_2 ratio. Results showed that the sensitivity to the amount of sulfur released to the gas phase as H_2S during devolatilization and concentration ratio of Na_2S and Na_2SO_4 in BLS have relatively large effects on performance response parameters as compared to carbon in volatile matter and CO/CO_2 concentration. The influential parameters appeared to be of great importance during model validation of CFD model against experimental data is considered.

The pressurized Chemrec BLG systems offer the following benefits when compared to recovery boilers:

- Effective pulping cooking chemicals recovery with simultaneous production of synthesis gas for high value-added green product streams.
- Increased energy efficiency from utilization of syngas as feedstock for production of green electricity, automotive fuels, or hydrogen.
- Dramatically improved green electricity yield through CC power generation in the Chemrec BLGCC system.
- Addition of new green chemical products, methanol, or DME for automotive fuels from the Chemrec BLGMF system.
- Potential of novel production of green hydrogen for fuel cells and other uses from the Chemrec BLGH2 system.
- Option of utilizing new kraft cooking processes with higher yield.
- New opportunity to manage the environmental impact of black liquor conversion.
- No risk of smelt—water explosion.

Figure 3.5 shows a Chemrec DP-1, and the new BioDME plant; Fig. 3.6 shows a block flow diagram of the BioDME project and Fig. 3.7 shows BioDME truck. The BioDME project aims to demonstrate production of environmentally optimized synthetic biofuel from lignocellulosic biomass at industrial scale. The project involves a consortium of Chemrec, Haldor Topsøe, Volvo, Preem, Total, Delphi, etc. The project is supported by the Swedish Energy Agency and the European Union's Seventh Framework Programme. The final output of this demonstration is DME produced from black liquor through the production of clean synthesis gas and a final fuel synthesis step. In order to check technical standards, commercial possibilities and engine compatibilities the BioDME will be tested in a fleet consisting of 14 Volvo trucks. In April 2012, Chemrec signed a cooperation agreement with China Tianchen Engineering Corporation (TCC) to provide an integrated offering of Chemrec plants on a global lump-sum turn-key basis and comarket the Chemrec BLG technology—a route to second generation biofuels or green power. The signing took place in Stockholm in the presence of the prime ministers of China (Wen Jiabao)

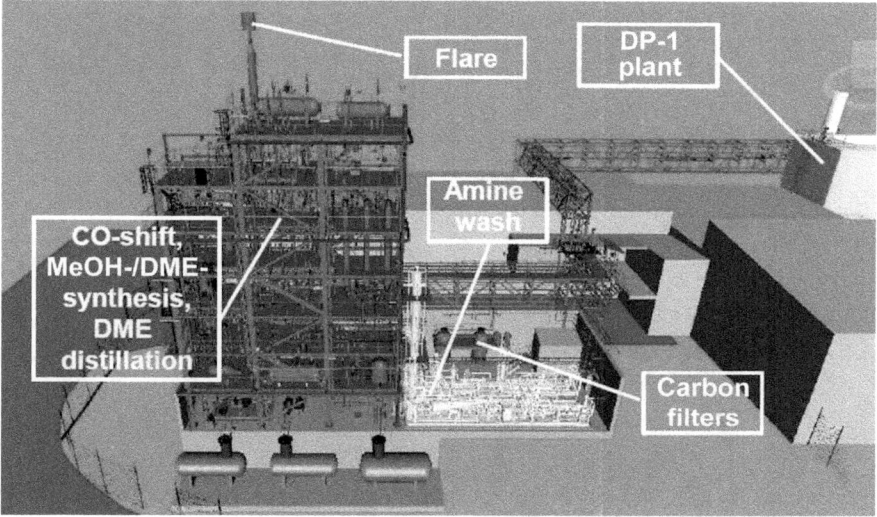

Figure 3.5 Chemrec DP-1 and the new BioDME plant. Reproduced with permission from Chemrec (2012).

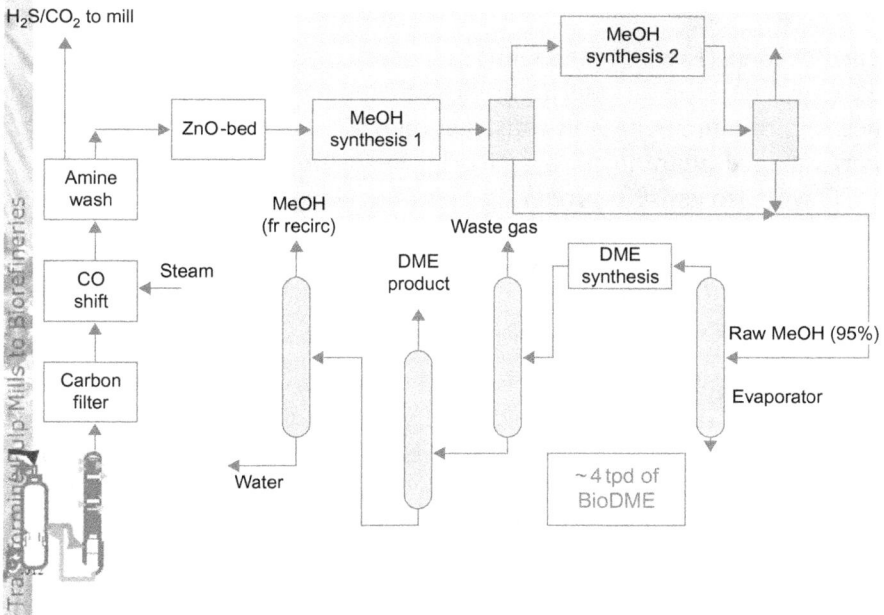

Figure 3.6 Block flow diagram of the BioDME project. Reproduced with permission from Chemrec (2012).

Customer: BDX Företagen AB
Truck supplying SmurfitKappa paper plant with timber

Figure 3.7 BioDME truck. Reproduced with permission from Chemrec (2012).

and Sweden (Fredrik Reinfeldt). Under the agreement, Chemrec and TCC will develop an offering to provide industry standard design, engineering, procurement, and construction (EPC) services as well as overall performance guarantees to support project financing for BLG plants. TCC will also assist in procuring plant financing.

The possible chemicals that can be produced from the syngas are hydrogen, methanol, DME, Fischer–Tropsch fuels, ethanol, and methyl tertiary butyl ether (MTBE) (Tampier et al., 2004).

The investment cost for a full-scaled pressurized black liquor gasification (PBLG) unit is estimated to be slightly higher than for a new conventional recovery boiler (Warnqvist et al., 2000). However, pressurized black liquor gasification with an integrated combined cycle (BLGCC) has the potential to double the amount of net electrical energy for a kraft pulp mill when compared to a modern recovery boiler with a steam turbine (Axegård, 1999). For more closed systems with less need of steam, this increase in electrical energy will be even higher. Another advantage with the PBLG process is the increased control of the fate of sulfur and sodium in the process that can be used to improve the pulp yield and the quality for the mill. This control is very important for the green liquor quality and is quite limited with a conventional recovery boiler. A disadvantage with gasification is that it will increase the causticizing load. However, BLG has a lower requirement for make-up salt cake compared to the recovery boiler. Even though the PBLG process might have a lot of advantages compared to the recovery boiler, there are still a number of uncertainties for this technology.

BLG is still a developing technology. Only small (100−350 tds/d) commercial atmospheric units have been built. Similar size pressurized demonstration units do not yet exist. It will take some time before reliable large units are available. BLG can produce more electricity (Vakkilainen et al., 2008). Current commercial atmospheric processes are not as energy efficient as the kraft recovery boiler process (Grace and Timmer, 1995; Mckeough, 2003). The black liquor gasifier needs to operate under pressure to have an electricity advantage.

Even though there are significant gains to be made, there are still remain many unresolved issues (Katofsky et al., 2003; Tucker, 2002): finding materials that survive in a gasifier, mitigating increased causticizing load, how to startup and shutdown, tar destruction, alkali removal, and achieving high reliability. The full impact of the BLG on recovery cycle chemistry needs to be carefully studied with commercial units. The first large demonstration units will cost 2−3 times more than a conventional recovery boiler. Although this will improve with time, price will hinder the progress of BLG. A small BLG with a commercial gas turbine size of 70 MWe requires a mill size of over 500,000 adt/a. Commercial gasifiers probably need to be over 250 MWe in size. It is therefore expected that full size black liquor gasifiers will be built in new greenfield mills and not as replacement units of old recovery boilers.

Clearly, BLG technology offers tremendous potential to make an impact on society. However, before it can totally replace the current recovery boiler technology, some work must be done to make it more economically attractive. One major area that requires attention is the causticization process. Gasification technology can cause significant increases in capacity for the lime cycle, requiring significant increases in fossil fuel consumption, and to improve economic viability, alternative causticization technologies must be considered.

Gasification is a well-established technique, but its application to black liquor is new and creates specific research needs. Perhaps the highest priority is to deal with the materials for constructing the gasifier. The process can operate at very high temperatures (up to 1,000°C) and involves very aggressive molten salts (Na_2S, Na_2CO_3, NaCl) that tend to react strongly with ceramics and other materials. There is a very aggressive gas atmosphere (HCl, CO). This was an issue with the gasification system at Weyerhaeuser, New Bern. The problem has now been solved by using new materials and making some design changes (Brown et al., 2004). There are issues concerning the formation of tar and condensable organic matter. Approximately 1−5% of the carbon in black liquor is converted to methanol, ethanol, cresol, xylene, and a variety of other tar and condensable organic components. Several other questions need to be addressed. For example, can sodium and sulfur separation be controlled by process design or operation? How much H_2S is produced, rather than other sulfur-containing gases? And can H_2S be recovered efficiently from the product gases? Researchers around the world are trying to find answers.

3.4 REMOVAL OF LIGNIN FROM BLACK LIQUOR

When annual pulp production is limited by recovery boiler capacity, it is possible to produce extra tons of pulp by separating lignin from black liquor. Precipitation of lignin from black liquor has been described earlier and has been described in the literature (Öhman, 2006). Lignin is a renewable material with an exciting future. From a short-term perspective (0−5 years), the most obvious use of the continuous bulk production of lignin is as a biofuel—by the pulp mill itself or by companies wanting to move away from fossil fuels. But there is also a good possibility that lignin will be much more valuable in the future. Lignin could very well become a globally-used base material with many different uses. Table 3.7 shows lignin content in black liquors of different woods. Figure 3.8 shows estimated black liquor world production.

Most of today's market pulp mills are self-sufficient in steam, from the black liquor alone, and have great potential to be energy suppliers to other industries and consumers. The energy surplus in mills can be exported in different ways, for example, as electricity, biofuels (bark and lignin), and heat for district heating. The most favorable

Table 3.7 Lignin in Black Liquors of Different Woods

	Lignin (kg/TP)
Spruce black liquor	510
Birch black liquor	340
Eucalyptus black liquor	340
Source: *Based on Hetemäki (2007).*	

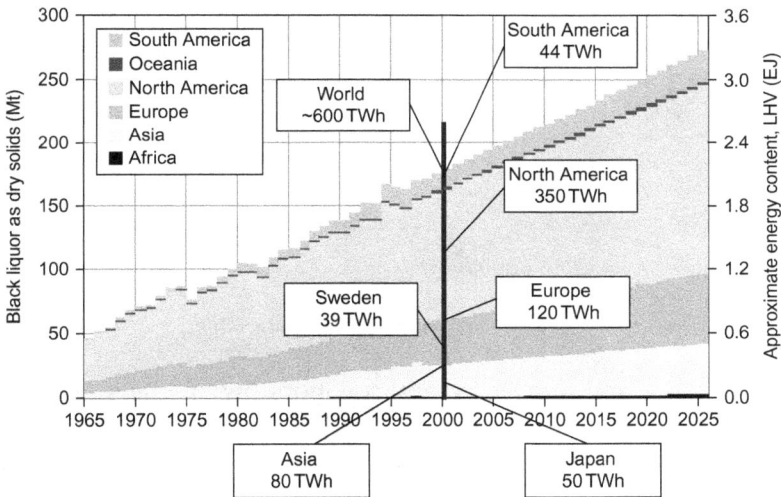

Figure 3.8 Estimated black liquor world production. Reproduced with permission from Landälv (2007).

alternative for each mill depends on its situation and has to be evaluated on a per-case basis. Lignin as a solid biofuel places certain demands on the design of the process system, just as all solid fuels do. A lot of different aspects, such as silo storage design, feeding into and out of the silo, drying of the product, and managing the explosion risks and moisture effects, must be taken into consideration.

LignoBoost is a complete process solution supplied by Metso (Fig. 3.9). A LignoBoost plant is roughly $25 \times 20 \times 14$ m^3 in size and includes all the equipment necessary for lignin extraction. LignoBoost has been developed through research cooperation between Innventia and Chalmers University of Technology. In May 2008, Innventia sold LignoBoost to Metso, which together with Innventia has refined the technology and developed a product that is now ready to enter the

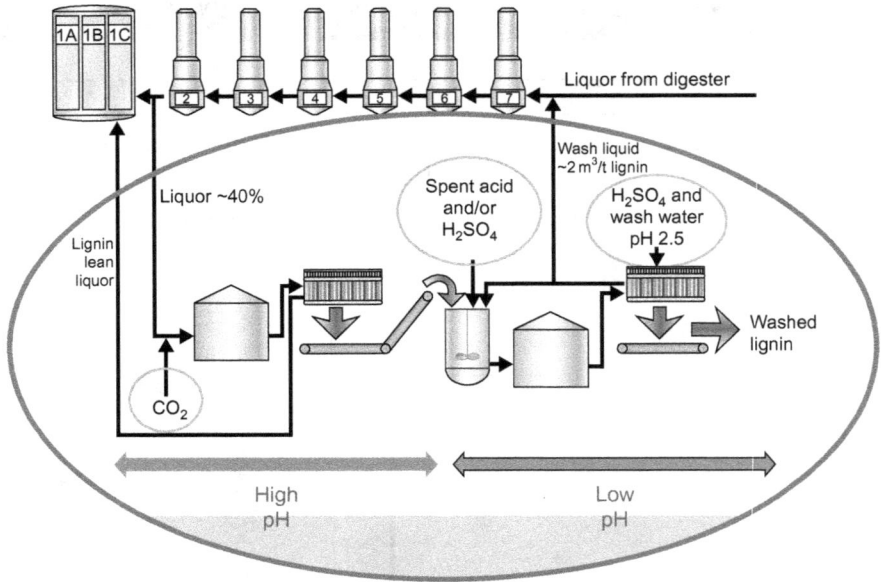

Figure 3.9 LignoBoost process. Reproduced with permission from Tomani (2010).

market. Metso and Innventia are collaborating in the research and development of LignoBoost and the use of lignin as biofuel. They see LignoBoost as an extremely important piece of the puzzle in the development of modern kraft pulp mills for biorefineries and producers of valuable new products. Lignin can be used as a high-quality biofuel, and it has the potential to be a biobased raw material for chemicals and materials.

There are four main operations in the LignoBoost process. These are: precipitation, dewatering, resuspension, and final washing. In the LignoBoost process, a stream of black liquor is taken from the black liquor evaporation plant, then lignin is precipitated by acidification, preferably with CO_2, and filtered. Instead of washing lignin immediately after filtration, as in traditional processes, the filter cake is redispersed and acidified. The resulting slurry is then filtered and washed by means of displacement washing. When the filter cake is redispersed in a liquid, at pH level and temperature values approximately equal to those of the final washing liquor, the concentration gradients during the washing stage will be low. The change in the pH level, most of the change in ionic strength, and any change in lignin solubility will then take place in the slurry, and not in the filter cake or in the filter medium during washing. The filtrate from chamber press filter is recycled to the black liquor evaporation plant, after the point at which the feed stream to LignoBoost is taken. This should avoid any decrease in lignin concentration in the stream fed to the LignoBoost operation. The filtrate from another chamber press filter (filtration, washing, and dewatering stages) is recycled to the weak black liquor. In some cases, this filtrate can be also used for washing the unbleached or oxygen delignified pulp. The LignoBoost process therefore makes it possible to extract lignin efficiently from the black liquor in kraft mill (Axegård, 2006a,b; 2007a,b; Axegard et al., 2007c; Neumann, 2008). LignoBoost process has been verified, in a demonstration plant located in Bäckhammar.

The LignoBoost process enables the fast production of high quality lignin at a low cost. Low-filtration resistances can be maintained throughout the process and an even lignin filter cake that is easy to wash and finally dewater, is formed in the second filtration/washing stage. Using the novel process, the specific filtration resistance is one to two orders of magnitude lower compared with the separation and washing made in a single filtration step. The separation of the pH and the ion strength reduction in two different steps results in the lignin becoming much more stable in all process stages with only a small amount of lignin dissolved during the final displacement washing.

Lignin coming directly from the LignoBoost plant has very good properties including 65−70% dry solids content, ash content of 0.1−0.5%, sodium 0.01−0.4%, and heating value of 26 GJ/t (Figs 3.10 and 3.11). It can be used as biofuel, replacing coal and oil, i.e., in pulp mill's power generation or in lime kilns. LignoBoost gives customers the possibility to increase the capacity of a pulp mill and turn pulp mills into significant energy suppliers. At the same, the extracted lignin is also of interest for other process industries as a raw material for plastics, coal fibers, and chemicals (Anon, 2007; Axegård, 2005, 2006a, b, 2007a,b; Axegard et al., 2007c; Frisell, 2008; Lennholm, 2007; Neumann, 2008; Tomani, 2010; Wallmo and Theliander, 2007).

Once processed, lignin works well as a fuel, as demonstrated by several tests. It has been used in burners together with bark and pulverized wood. Södra Cell Mönsterås pulp mill in Sweden, for instance, carried out a full-scale trial where 37 t of lignin was burnt as the lime kiln fuel. At one point in the three-day trial, lignin was the only fuel

65–70% DS
HHV (dry ash free): 26–27 MJ/kg

C: 63–66%
H: 5.7–6.2%
O: 26–27.5%
S: 1.8–3.2% ←
N: 0.1–0.2%

Ash (dry): 0.2–1.4%
Na: 120–230 g/kg ash
K: 25–80 g/kg ash

Figure 3.10 Typical LignoBoost softwood lignin from the demo plant. Reproduced with permission from Tomani (2010).

LingoBoost, a subsidary company of STFI-Packfrosk

(Jointly financed by STFI-Packforsk, Södra, Stora Enso, Fortum, Värme, and the Swedish Energy Agency)

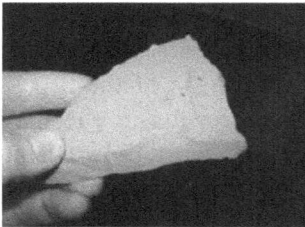

⟶ Electricity + heat

High quality lignin extracted from black liquor (kraft pulp mill)

Pellets directly from the pieces of a kraft lignin cake

Could also produce chemicals

Figure 3.11 Processing lignin to electricity and heat in Sweden. Reproduced with permission from Hetemäki (2007).

being used. Another trial was carried out at Fortum's Värtaverket heat and power plant in Sweden. In this trial, 4,000 t of lignin was used in four campaigns over the course of 13 weeks. During eight of these weeks, 8–15% of the total energy output came from lignin instead of coal. However, estimates indicate that it should be possible to mix more than 30% lignin into the fuel with optimized equipment.

Lignin can also be mixed into liquid fuels, like fossil fuel oil (to at least increase the amount of biofuel used), tall oil pitch, or other liquid fuels. There are limits as to how much lignin can be mixed into these types of solid/liquid slurries, but using only simple equipment it has been possible to mix as much as 45% lignin into fuel oil or tall oil pitch.

Using lignin as fuel is only the first step. Lignin can also be refined into a wide range of products. It has been transformed into low-cost carbon fiber, which has a strong potential to grow in demand as new vehicles will need to be lighter in order to consume less fuel. Activated carbon is another product with potential to be made from lignin. Stricter requirements on emissions of, for example, heavy metals will increase the demand for activated carbon as well. The plastics industry is another huge market where lignin could be used as a base chemical. One example here is phenols or mixes of phenols. Global production of phenols in 2006 was 8,000,000 t. Phenols are currently made from fossil substances. Investing in lignin extraction will have a direct positive effect on pulp production. Lignin also produces a renewable fuel that in the future has the potential to become a base substance with a wide range of uses and a global market.

Lignins are used as binders, dispersants, emulsifiers, and sequestrants. It has been proposed to isolate phenols from lignin and to produce carbon fibers (Griffith et al., 2003; Kadla et al., 2002). The LignoBoost process also offers new opportunities for further use of a kraft pulp mill as a biorefinery such as in xylan removal from black liquor, biomass gasification, and ethanol fermentation (Axegard, 2007a, b; Rodden, 2007).

Precipitation of lignin requires carbon dioxide. The bulk of the variable cost is due to carbon dioxide if commercial product is used. It may be possible to use carbon dioxide from the lime kiln but gas cleaning is a challenge. Carbon dioxide from ethanol fermentation yields about 1 t of pure carbon dioxide per ton of ethanol produced. Currently, sized ethanol plants are too small to justify recovery of the produced carbon dioxide. By combining lignin production with ethanol production, the carbon dioxide can efficiently be utilized and the economical performance significantly improved.

The amount of lignin (and xylan) that can be removed from black liquor depends mainly on the status of the recovery boiler. At a certain amount of heat value in the fired black liquor, the performance is deteriorated. In many mills, this critical level is between 10% and 30% of lignin removed. One interesting way to handle this is to add fuel gas from gasified biomass and thus compensate for lost heat value. Produced carbon dioxide can also be used for lignin precipitation (Axegård, 2006b). The ultimate development would be removal of all valuable organic components from the black liquor such as lignin, xylan, and sugar acids and instead obtain all the fuel need from gasified biomass such as forestry residuals. Such an approach would make a complete removal of organic components in black liquor possible. The traditional recovery boiler may also be replaced with less capital-demanding, and less complicated, techniques.

Metso will supply the world's first commercial installation of LignoBoost technology to Domtar in North America. The equipment will be integrated with the Plymouth North Carolina pulp mill. The LignoBoost process separates and collects lignin from pulping liquor. This order is an important breakthrough for Metso's patented LignoBoost technology and provides the Plymouth North Carolina mill with numerous benefits. Separation of a portion of the mill's total lignin production off-loads the recovery boiler and allows an increase in pulp production capacity. The lignin recovered will be used for internal and external applications. This project is a potential game changer for the pulp and paper industry because it will allow pulp mills to have a new, more profitable, value stream from a product that was traditionally burned in a recovery boiler. Domtar Corporation is the largest integrated manufacturer and marketer of uncoated freesheet paper in North America, the second largest in the world based on production capacity and is also a manufacturer of papergrade, fluff, and specialty pulp. The company designs, manufactures, markets, and distributes a wide range of business, commercial printing, and publishing as well as converting and specialty papers. Domtar also produces a complete line of incontinence care products and distributes washcloths marketed primarily under the Attends® brand name. Domtar owns and operates ArivaTM, an extensive network of strategically located paper distribution facilities.

The LignoBoost project is based on the 24 t/d demonstration plant in Bäckhammar in Sweden (Fig. 3.12). The Bäckhammar plant is owned and operated by Innventia since 2006. In 2008, Metso acquired the LignoBoost technology from Innventia and the companies have

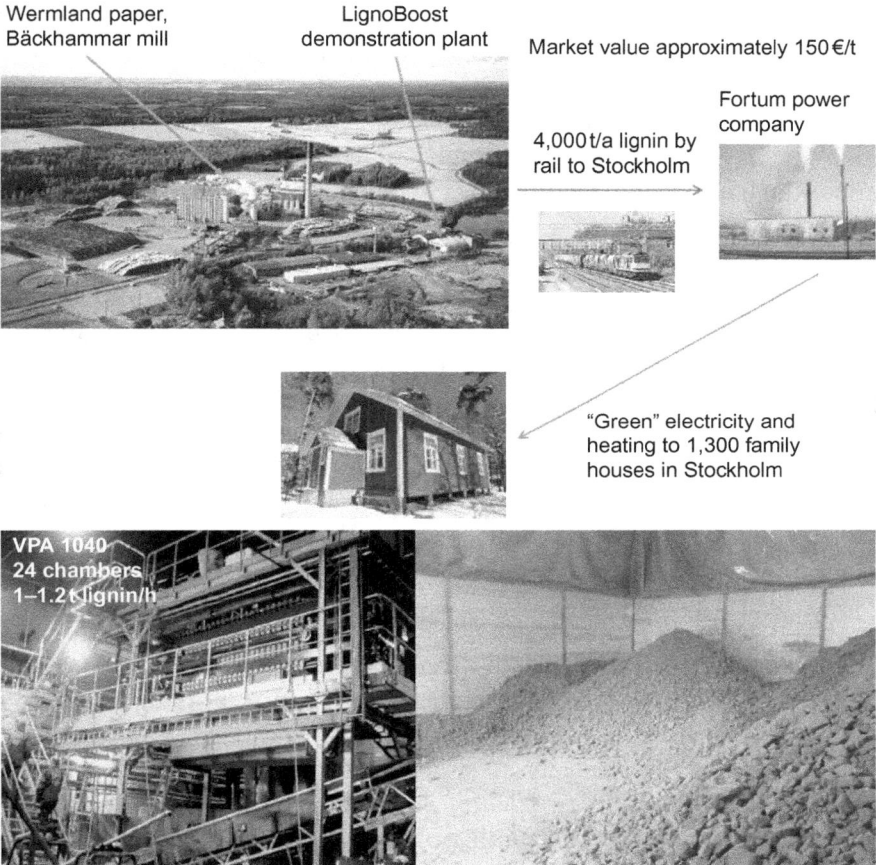

Wermland paper, Bäckhammar mill

LignoBoost demonstration plant

Market value approximately 150 €/t

4,000 t/a lignin by rail to Stockholm

Fortum power company

"Green" electricity and heating to 1,300 family houses in Stockholm

VPA 1040 24 chambers 1–1.2 t lignin/h

Figure 3.12 LignoBoost demonstration plant in operation since late 2006. Reproduced with permission from Tomani (2010); Hetemäki (2007).

since then been working together on the commercialization of the process. The LignoBoost plant is expected to be in commercial operation in early 2013.

3.5 OTHER PRODUCTS

Typically, the tall oil produced in a pulp and paper mill is burnt in the recovery boiler. Instead it could be used to produce green fuels, e.g., biodiesel. Producing biodiesel from tall oil via hydrogenation is more economically attractive than producing biodiesel via esterification (Prakash, 1998). This process could make biodiesel from tall oil competitive with traditionally derived diesel. In some countries, most crude tall oil is currently incinerated as fuel in the lime kilns of pulp mills to displace fossil fuel, and in some countries, where extractive content of the wood is much higher, tall oil is fractionated into the crude tall oil into value-added components. The processing of tall oil into a high quality diesel additive has been researched in the laboratory and pilot scale. The later studies included promising road tests by Canada Post Corporation (Ragauskas, 2006). Given that many kraft pulp mills already collect these extractives, their future utilization for fuels will be based on competing economic considerations. Fatty acids can be directly esterified by alcohols into diesel fuel, while the rosin acids can be converted by the "Super Cetane" hydrogenation process developed in Canada. Turpentine recovered from process condensates in Canadian mills is generally incinerated as fuel in one of the onsite boilers. Processing it into consumer grade products is possible but, in many cases, it is more valuable as a fuel. Thorp (2005b) has reported production-rate potential of 530 million liters diesel per year in the USA.

The average thousand-tons-per-day softwood kraft mill has approximately 7 t/d of methanol in its foul condensate streams. Most mills use steam strippers to concentrate the methanol to about half its volume before incineration. Some mills use air strippers, which do not remove methanol effectively or simply send foul condensates to effluent treatment where the methanol is consumed by biological activity. It is possible to purify this methanol for alternative uses, either onsite or for sale. One pilot project has used the catalytic conversion process for converting the methanol to formaldehyde. Methanol is an important industrial chemical, and one that can be used to produce a wide range of other chemicals. Roughly 35–40% of methanol is used to create formaldehyde as a feedstock for phenolic resins, much of which ultimately is used as an adhesive in plywood. Additionally, roughly 20% is used in gasoline, either directly or as MTBE. The remaining production is used to create substances such as acetic acid. Methanol and hydrogen have been rated the best near-term prospects for bio-based syngas products because other products have been found to require additional development and are not currently economically viable. The methanol market is mature. However, methanol prices have fallen due to the economic slowdown. The primary uses of methanol are both impacted by the slowdown. For instance, the housing market represents roughly half of formaldehyde demand in North America.

The housing market has weakened and the particle board, fiber-board, and similar industries in North America are under stress due to international competition. Additionally, MTBE has been under pressure in many countries due to mandates to use renewable ethanol-based Ethyl tert-butyl ether (ETBE) or ethanol in fuels.

Waste organics sent to effluent treatment at pulp and paper mills are unique compared with municipal organic wastes, which have a very high carbon-to-nitrogen ratio. Certain bacteria in activated sludge treatment systems under such conditions accumulate 3-hydroxybutyric acid (PHB), a potential building block for biopolymers. Extraction of PHB remains the significant hurdle to this process. Pulp and paper waste treatment sludge is typically buried in landfills, incinerated or spread on land as a nutrient enhancer. Research is under way to improve the performance of microbes in the conversion of nutrients in effluents to PHB and other fermentation products.

The USA pulp and paper industry processes 108 million tons of pulpwood per annum. At least 14 million tons of hemicellulose (2 billion gallons ethanol; 600 million gallons acetic acid; $3.3 billion net cash flow), 5 million tons of paper mill sludge (feedstock for ethanol; no pretreatment), and 700 million liters of turpentine and tall oil (feedstock for biodiesel) per annum is available.

In an optimized forest biorefinery, part of the hemicellulose that is now burned would be used to create new, more valuable products (Thorp, 2005a,b; Thorp and Raymond, 2005; Thorp et al., 2008). A portion of hemicellulose can be extracted from wood chips prior to pulping using hot water extraction in low-pressure digesters. Some acetic acid is formed during the extraction process and this must be separated from the sugar solution. The sugars can then be fermented to ethanol or other high-value chemicals, creating an additional product stream. Removing part of the hemicellulose prior to the digester will increase the throughput potential of the pulping process. However, utilizing some of the hemicellulose as a sugar feedstock reduces the energy content of the pulping byproduct black liquor, which is an important renewable energy source for kraft pulp mills. In the future, to fully optimize the forest biorefinery, the economic and energy implications of diverting a portion of hemicellulose to other products will need to be balanced. The loss of this energy source can be offset by improved energy efficiency in the pulp and paper manufacturing process. Ultimately, forest biorefineries would potentially use a combination of new technologies that result in more complete, energy efficient, and cost-effective use of the wood feedstock (Chambost and Stuart, 2007; Closset, 2004; Connor, 2007).

REFERENCES

Ådahl, A., Harvey, S., Berntsson, T., 2004. Process industry energy retrofits: the importance of emission baselines for greenhouse gas reductions. Energy Policy 32, 1375–1388.

Ali, O.F., Cenicola, J.T., Li, J., Taylor, J.D., 2005. Process for producing alkaline treated cellulosic fibers, US Patent 6,896,810.

Anon, 2007. LignoBoost does business with lignin fuel. Beyond 2, 4–5.

Axegård, P., 1999. Kretsloppsanpassad massafabrik-Slutrapport, KAM 1 1996–1999, KAMrapport A31, Stiftelsen för Miljöstrategisk forskning.

Axegård, P., 2005. The future pulp mill—a biorefinery. First International Biorefinery Workshop, July 20–21, 2005, Washington, DC.

Axegård, P., 2006a. Lignin removal from black liquor for increased energy efficiency and pulp capacity increase. Energy Management for Pulp and Papermakers. 16–18 October, Budapest, Hungary, Paper 12, pp. 31.

Axegård, P., 2006b. Utilization of black liquor and forestry residues in a pulp mill biorefinery. Forest Based Sector Technology Platform Conference. 22–23 November, Lahti, Finland.

Axegård, P., 2007a. Lignin from black liquor: a valuable fuel and chemical feedstock. Biorefining for the Pulp and Paper Industry. 10–11 December, Stockholm, Sweden, pp. 34.

Axegård, P., 2007b. The kraft pulp mill as a biorefinery. Third ICEP International Colloquium on Eucalyptus Pulp. 4–7 March, Belo Horizonte, Brazil, pp. 6.

Axegård, P., Backlund, B., Tomani, P., 2007c. The pulp mill based biorefinery. Pulp Paper 2007 Conference. Biomass Conversions. 5–7 June, Helsinki, Finland, pp. 19–26.

Bajpai, P., 2008. Chemical Recovery in Pulp and Paper Making. PIRA International, UK, p. 166.

Bajpai, P., 2012. Biotechnology for Pulp and Paper Processing. Springer-Verlag Inc., New York, NY.

Bozell, J.J., Black. S.K., Myers. M., 1995. Clean fractionation of lignocellulosics—a new process for preparation of alternative feedstocks for the chemical industry. Eighth International Symposium on Wood and Pulping Chemistry, Helsinki, Finland, 6–9 June, pp. 697–704.

Brown, C., Landälv, I., 2001. The Chemrec black liquor recovery technology—a status report. International Chemical Recovery Conference. 11–14 June, Whistler, Canada.

Brown, C.A., Gorog, J.P., Leary, R., Abdullah, Z., 2004. The Chemrec black liquor gasifier at new bern—a status report. International Chemical Recovery Conference. 6–10 June, Charleston, SC, USA.

Brownell, H.H., Saddler, J.N., 1987. Steam pretreatment of lignocellulosic material for enhanced enzymatic hydrolysis. Biotechnol. Bioeng. 29, 228–235.

Cara, C., Ruiz, E., Ballesteros, I., Negro, M.J., Castro, E., 2006. Enhanced enzymatic hydrolysis of olive tree wood by steam explosion and alkaline peroxide delignification. Process Biochem. 41, 423–429.

Chambost, V., Stuart, P.R., 2007. Selecting the most appropriate products for the forest biorefinery. Ind. Biotechnol. 3 (2), 112–119.

Chemrec gasification technology—turns pulp and paper mills into bio-refineries, 2009 <www .chemrec.se> (accessed 10.08.2012).

Chemrec, 2012. Personal communication.

Christopher, L., 2012. Adding value prior to pulping: bioproducts from hemicellulose. Global Perspectives on Sustainable Forest Management book edited by Okia Clement Akais ISBN 978-953-51-0569-5 doi:10.5772/36849.

Closset, G., 2004. Advancing the forest biorefinery. Forest Products Techno-Business Forum. 26–27 October, Atlanta, GA.

Connor, E., 2007. The integrated forest biorefinery: the pathway to our bio-future. International Chemical Recovery Conference: Efficiency and Energy Management. 29 May–1 June, Quebec City, QC, Canada, pp. 323–327.

Cruz, J.M., Dominguez, J.M., Dominguez, H., Parajo, J.C., 1999. Solvent extraction of hemicellulosic wood hydrolyzates: a procedure useful for obtaining both detoxified fermentation media and polyphenols with antioxidant activity. Food Chem. 67, 147–153.

Cunningham, R.L., Carr, M.E., Bagby, M.O., 1986. Hemicellulose isolation of annual plants. Biotechnology and Bioengineering. Symposium, No 17, Symposium on Biotechnology for Fuels and Chemicals, (Gatlinburg) 8th: 159–168 (May 13–16, 1986).

Dahlquist, E., 2003. A combined physical and statistical simulation model for black liquor gasification. 44th Conference on Simulation and Modeling on September 18 -19, 2003 in Västerås, Sweden.

Dahlquist, E., Jacobs, R., 1994. Development of a dry black liquor gasification process. Pulp Pap. Can. 95, 2.

Dahlquist, E., Avelin, A., Yan, J., 2009. Black liquor gasification in a CFB gasifier—system solutions. First International Conference on Applied Energy (ICAE09). 5–7 January 2009, Hong Kong.

DeCarrera, R., 2006. Quarterly Technical Progress Report 20 Demonstration of Black Liquor Gasification at Big Island, Report 40850R20 <http://www.gp.com/containerboard/mills/big/pdf/ rpt40850R20.pdf> (accessed 20.10.2012).

Durai-Swamy, K., Mansour, M.N., Warren, D.W., 1991. Pulsed combustion process for black liquor gasification, US DOE Report DOE/CE/40893-T1 (DE92003672).

Eckert, C.A., Bush, D., Brown, J.S., Liotta, C.L., 2000. Tuning solvents for sustainable technology. Ind. Eng. Chem. Res. 39 (12), 4615–4621.

Eckert, C.A., Liotta, C.L., Bush, D., Brown, J., Hallett, J., 2004. Sustainable reactions in tunable solvents. J. Phys. Chem. B 108, 18108–18118.

Ebringerova, A., Hromadova, Z., Kaucurakova, M., Antal, M., 1994. Quaternized xylans: synthesis and structural characterization. Carbohyd. Polym. 24, 301–308.

Ekbom, T., Lindblom, M., Berglin, N., Ahlvik, P., 2003. Technical and commercial feasibility study of black liquor gasification with methanol/DME production as motor fuels for automotive uses—BLGMF. Report for Contract No. 4.1030/Z/01-087/2001, European Commission, Altener program, Stockholm, Sweden.

Farmer, M.C., 2005. The adaptable integrated biorefinery for existing pulp mills. TAPPI Engineering, Pulping, and Environmental Conference. 28–31 August, Philadelphia, PA.

Farmer, M., Sinquefield, S., 2003. An External Benefits Study of Black Liquor Gasification. Georgia Institute of Technology (Final Report, June 15, 2003).

Frisell, H., 2008. Breakthrough for new Swedish environmental technology. Dagens Ind. 33 (69), 26, 25 March.

Fitzpatrick, S.W., 1997. US Patent 5,608,105.

Gabrielii, I., Gatenholm, P., Glasser, W.G., Jain, R.K., Kenne, L., 2000. Separation, characterization and hydrogel-formation of hemicellulose from aspen wood. Carbohyd. Polym. 43, 367–374.

Gonzalez, J., Cruz, J.M., Dominguez, H., Parajo, J.C., 2004. Production of antioxidants from *Eucalyptus globulus* wood by solvent extraction of hemicellulose hydrolyzates. Food Chem. 84, 243–251.

Griffith, W.L., Compere, A.L., Leitten, C.F., Shaffer, J.T., 2003. Low-cost, lignin-based carbon fiber for transportation applications. Int. SAMPE Tech. Conf. 35, 142–149.

Grace, T., 1987. Chemical recovery technology—a review. IPC Technical Paper series No. 247.

Grace, T.M., 1992. Chemical recovery process chemistry, Chemical Recovery in the Alkaline Pulping Processes, third ed. TAPPI Press, Atlanta, GA, pp. 57–78.

Grace, T.M., Timmer, W.M., 1995. A comparison of alternative black liquor recovery technologies. Proc. Int. Chem. Rec. Conf., Toronto, B269–275.

Harvey, S., Facchini, B., 2004. Predicting black liquor gasification combined cycle powerhouse performance accounting for off-design gas turbine operation. Appl. Therm. Eng. 24, 111–126.

Hashimoto, T., Hashimoto, K., 1975. Studies on the utilization of xylan and glucomannan in woods. I. Purification and separation. Yakugaku Zasshi 95 (10), 1239–1244.

Heitz, M., Carrasco, F., Rubio, M., Chauvette, G., Chornet, E., Julian, L., et al., 1986. Generalised correlations for the aqueous liquefaction of lignocellulosics. Can. J. Chem. Eng. 64, 647–650.

Heitz, M., Capek-Menard, E., Koeberle, P.G., Gagne, J., Chornet, E., Overend, R.P., et al., 1991. Fractionation of *Populus tremuloides* in the pilot plant scale: optimization of steam pretreatment conditions using STAKE II technology. Bioresour. Technol. 35, 23–32.

Hetemäki, L., 2007. Forest biorefineries: current status and outlook. IUFRO Division VI Symposium. 14–20 August, Saariselkä, Finland.

Hsu, T.A., 1996. Pretreatment of biomass. In: Hsu, T.-A., Wyman, C.E. (Eds.), Handbook on Bioethanol Production and Utilization, Applied Energy Technology Series. Taylor & Francis, Washington, DC, p. 1996. (Chapter 10).

Huang, H.J., Ramaswamy, S., Tschirner, U.W., Ramarao, B.V., 2008. A review of separation technologies in current and future biorefineries. Sep. Purif. Technol. 62, 1–21.

Hupa, M., Backman, R., Frederick, W.J., 1994. Black liquor combustion properties. International Conference Baltic Sea, 24–26 May, Finnish Recovery Boiler Committee.

Ibrahim, M., Glasser, W.G., 1999. Steam-assisted biomass fractionation. Part III: a quantitative evaluation of the "clean fractionation" concept. Bioresour. Technol. 70, 181–192.

Jain, R.K., Sjostedt, M., Glasser, W.G., 2000. Thermoplastic xylan derivatives with propylene oxide. Cellulose 7 (4), 319–336.

Josefsson, T., Lennholm, H., Gellerstedt, G., 2002. Steam explosion of aspen wood. Characterisation of reaction products. Holzforschung 56, 289–297.

Katofsky, R., Consonni, S., Larson, E.D., 2003. A cost–benefit analysis of black liquor gasification combined cycle systems. Proceedings, TAPPI Fall Technical Conference: Engineering, Pulping and PCE&I, Chicago. 26–30 October, p 22.

Kadla, J.F., Kubo, S., Venditti, R.A., Gilbert, R.D., Compere, A.L., Griffith, W., 2002. Lignin-based carbon fibers for composite fiber applications. Carbon 40, 2913–2920.

Kignell, J.E., 1989. Process for chemicals and energy recovery from waste liquors, US Patent 4,808,264.

Knappert, D.R., Grethlein, H.E., 1981. Converse AO, partial acid hydrolysis of poplar wood as a pretreatment for enzymatic hydrolysis. Biotechnol. Bioeng. 11, 67–77.

Kubikova, J., Zemann, A., Krkoska, P., Bobleter, O., 1996. Hydrothermal pretreatment of wheat straw for the production of pulp and paper. TAPPI J. 79, 163–169.

Kulkarni, A., Mathur, R., Naithani, S., Pant, R., 2007. Direct Alkali Regeneration System (DARS) in Small Pulp Mills. Central Pulp and Paper Research Institute, Dehra Dun, India.

Landälv, I., 2007. The status of the Chemrec black liquor gasification concept. Second European Summer School on Renewable Motor Fuels. 29–31 August, Warsaw, Poland.

Landälv, I., 2012. Personal communication.

Larsen, E., Kreutz, T., Consonni, S., 1998. Performance and preliminary economics of black liquor combined cycles for a range of kraft pulp mill sizes. International Chemical Recovery Conference. 1–4 June, Tampa, FL, USA, vol. 2. pp. 675–692.

Larsen, E.D., McDonald, G.W., Yang, W., Frederick, W.J., Iisa, K., Kreutz, T.G., et al., 2000. A cost–benefit assessment of BLGCC technology. TAPPI J. 83 (6), 1–15.

Larsen, E., Consonni, S., Katofsky, R., 2003. A Cost–Benefit Assessment of Biomass Gasification Power Generation in the Pulp and Paper Industry. Princeton Environmental Institute (Final Report, October 8, 2003), Princeton, NJ.

Larsen, E., Consonni, S., Katofsky, R., 2006a. A Cost–Benefit Assessment of Gasification-Based Biorefining in the Kraft Pulp and Paper Industry, vol. 1. Princeton University and Politecnico di Milano, Final Report.

Larsson, A., Nordin, A., Backman, R., Warnqvist, B., Eriksson, G., 2006b. Influence of black liquor variability, combustion, and gasification process variables and inaccuracies in thermochemical data on equilibrium modeling results. Energy Fuels 20, 359–363.

Lazzaroni, M.J., Bush, D., Brown, J.S., Eckert, C.A., 2005. High pressure vapor and liquid equilibria of some carbon dioxide and organic binary systems. J. Chem. Eng. Data 50 (1), 60–65.

Lee, Y.Y., Iyer, P., Torget, R.W., 1999. Dilute-acid hydrolysis of lignocellulosic biomass 65, pp. 93–115.

Lennholm, B., 2007. Lignin from the pulp mills' black liquor: new biofuel with promising potential. Nord. Papperstidn. 6 (June), 16.

Lesutis, H.P., Gläser, R., Griffith, K., Liotta, C.L., Eckert, C.A., 2001. Near critical water: a benign medium for catalytic reactions. Ind. Eng. Chem. Res. 40, 6063–6067.

Li, X., Simonsen, J., Li, K., 2004. Wood dissolution and the regeneration of its components using ionic liquids. 227th American Chemical Society National Meeting Abstracts. Anaheim, California, 28 March – 1 April, 2004.

Lindblom, M., 2003. An overview of Chemrec process concepts. Sixth International Colloquium on Black Liquor Combustion and Gasification. 13–16 May, Park City, Utah, USA.

Lindblom, M., 2006. Chemrec pressurized black liquor gasification—status and future plans. Seventh International Colloquium on Black Liquor Combustion and Gasification. 31 July–2 August, Jyväskylä, Finland.

Lora, J.H., Wayman, M., 1978. Delignification of hardwoods by autohydrolysis and extraction. TAPPI J. 61, 47–50.

Lu, J., Lazzaroni, M.J., Hallett, J.P., Bommarius, A.S., Liotta, C.L., Eckert, C.A., 2004. Tunable solvents for homogeneous catalyst recycle. Ind. Eng. Chem. Res. 43 (7), 1586–1590.

Lundqvist, J., Jacobs, A., Palm, M., Zacchi, G., Dahlman, O., Stålbrand, H., 2002. Characterization of galactoglucomannan extracted from spruce (*Picea abies*) by heat-fractionation at different conditions. Carbohyd. Polym. 51 (2), 203–211.

Mansour, M.N., Steedman, W.G., Durai-Swamy, K., Kazares, R.E., Raman, T.V., 1992. Chemical and energy recovery from black liquor by steam reforming. International Chemical Recovery Conference. 7–11 June, Seattle, WA, USA.

Mansour, M.N., Durai-Swamy, K., Aghamohammadi, B., 1993. Pulsed combustion process for black liquor gasification. Second Annual Report US DOE Report DOE/CE/40893-T2 (DE94002668).

Mansour, M.N., Durai-Swamy, K., Warren, D.W., 1997. Endothermic spent liquor recovery process, US Patent 5,637,192.

Marinova, M., Mateos-Espejel, E., Paris, J., 2010a. Successful conversion of a kraft pulp mill into a forest biorefinery: energy analysis issues. Twenty-Third ECOS, Paper 91, Lausanne.

Marinova, M., Eilers, H., Barreto Do Carmo, C., Paris, J., 2010b. Opportunity for furfural production from hardwood chips pre-hydrolysate. Third International IUPAC Conference on Green Chemistry, Ottawa, 15−18 August 2010.

Marklund, M., Tegman, R., Gebart, R., 2007. CFD modelling of BLG: identification of important model parameters. Fuel 86, 1918−1926 (ISSN 0016-2361).

Martin, N., Anglani, N., Einstein, D., Khrushch, M., Worrell, E., Price, L.K., 2000. Opportunities to Improve Energy Efficiency and Reduce Greenhouse Gas Emissions in the U.S. Pulp and Paper Industry. Ernest O. Lawrence Berkeley National Laboratory (Report, July 2000).

Mateos-Espejel, E., Moshkelani, M., Keshtkar, M., Paris, J., 2011. Sustainability of the green integrated forest biorefinery: a question of energy. J. Sci. Technol. For. Prod. Processes 1 (1), 55.

Mckeough, P., 2003. Evaluation of potential improvements to BLG technology. Colloquium of Black Liquor Combustion and Gasification. Park City, Utah, p. 12.

Menon, V., Prakash, G., Rao, M., 2010. Value added products from hemicelluloses: biotechnological perspective. Glob. J. Biochem. 1 (1), 36−67.

Middleton, T., 2006. Steam reforming technology at the Norampac Trenton mil. IEA Meeting, Annex XV Black Liquor Gasification. 20−22 February, Washington, NC.

Mok, W.S.L., Antal, M.J.J., 1992. Uncatalyzed solvolysis of whole biomass hemicellulose by hot compressed liquid water. Ind. Eng. Chem. Res. 31, 1157.

Molin, U., Teder, A., 2002. Importance of cellulose/hemicellulose-ratio for pulp strength. Nord. Pulp Pap. Res. 17 (1), 14−19, 28.

Moens, L., Khan, N., 2003. Application of room-temperature ionic liquids to the chemical processing of biomass-derived feedstocks. NATO Science Series, II. Math. Phys. Chem. 92, 157−171.

Möllersten, K., Yan, J., Westermark, M., 2003a. Potential and cost-effectiveness of CO_2-reduction in the Swedish pulp and paper sector. Energy 28, 691−710.

Möllersten, K., Yan, J., Moreira, J., 2003b. Potential market niches for biomass energy with CO_2 capture and storage. Biomass Bioenergy 25 (3), 273−285.

Mosier, N., Wyman, C., Dale, B., Elander, R., Lee, Y.Y., Holtzapple, M., et al., 2005. Features of promising technologies for pretreatment of lignocellulosic biomass. Bioresour. Technol. 96, 673−686.

N'Diaye, S., Rigal, L., 2000. Factors influencing the alkaline extraction of poplar hemicelluloses in a twin-screw reactor: correlation with specific mechanical energy and residence time distribution of the liquid phase. Bioresour. Technol. 75, 13−18.

N'Diaye, S., Rigal, L., Larocque, P., Vidal, P.F., 1996. Extraction of hemicelluloses from poplar, *Populus tremuloides*, using an extruder-type twin-screw reactor: a feasibility study. Bioresour. Technol. 57, 61−67.

Neumann, M., 2008. New uses for lignin in the biorefinery of the future. Nord. Papp. Massa 1, 42−43.

Newport, D.G., Rockvam, L., Rowbotton, R., 2004. Black liquor steam reformer start-up at Norainpac. Proceedings of TAPPI International Chemical Recovery Conference, South Carolina, 6−10, June 2004.

Nguyen, Q.A., Tucker, M.P., Keller, F.A., Eddy, F.P., 2000. Two stage dilute-acid pretreatment of softwoods. Appl. Biochem. Biotechnol. 84/86, 561−576.

Nilsson, L.J., Larson, E.D., Gilbreath, K.R., Gupta, A., 1995. Energy Efficiency and the Pulp and Paper Industry. ACEEE, Washington, DC/Berkeley, CA.

Niu, W., Molefe, M.N., Frost, J.W., 2003. Microbial synthesis of the energetic material precursor 1,2,4-butanetriol. J. Am. Chem. Soc. 125, 12998.

Nohlgren, I., 2004. Non-conventional caustization technologies: a review. Nord. Pulp Pap. Res. J. 19 (4), 467−477.

Nohlgren, I., Sinquefield, S., 2004. Black liquor gasification with direct causticization using titanates: equilibrium calculations. Ind. Eng. Chem. Res. 43 (19), 5996−6000.

Nolen, S.A., Liotta, C.L., Eckert, C.A., Gläser, R., 2003. The catalytic opportunities of near-critical water: a benign medium for conventionally acid and base catalyzed organic synthesis. Green Chem., 663−669.

Öhman, F., 2006. Precipitation and separation of lignin from kraft black liquor. PhD thesis, Chalmers Technical University, Gothenburg, Sweden.

Page, D.H., Seth, R.S., 1985. Strength and chemical composition of wood pulp fibers. The Eighth Fundamental Research Symposium. Oxford, UK, 15−20, September, pp. 77−91.

Palm, M., Zacchi, G., 2003. Extraction of hemicellulosic oligosaccharides from spruce using microwave oven or steam treatment. Biomacromolecules 4 (3), 617−623.

Patrick, K., Siedel, B., 2003. Gasification edges closer to commercial reality with three new N.A. mills startups. PaperAge October.

Perez, D.D., Huber, P., Petit-Conil, M., 2011. Extraction of hemicelluloses from wood chips and some examples of usage in the papermaking process. Colloque Interfibres. September 6−8, 2011, Bordeaux, France.

Prakash, C.B., 1998. A Critical Review of Biodiesel as a Transportation Fuel in Canada. Global Change Strategies International Inc., Ottawa, Canada, Report.

Ragauskas, A.J., Nagy, M., Kim, D.H., Eckert, C.A., Hallett, J.P., Liotta, C.L., 2006. From wood to fuels: integrating biofuels and pulp production. Ind. Biotechnol. 2 (1), 55−65.

Richards, T., Nohlgren, I., Warnqvist, B., Theliander, H., 2002. Mass and energy balances for a conventional recovery cycle and for a cycle using borate or titanates. Nord. Pulp Pap. Res. J. 17 (3), 213−222.

Rockvam, L.N., 2001. Black liquor steam reforming and recovery commercialization. International Chemical Recovery Conference. 11−14 June, Whistler, Canada.

Rodden, G., 2007. LignoBoost is proving its worth: Wermland paper is in the forefront of biofuel development thanks to an agreement with STFI-Packforsk. Pulp Pap. Int. 49 (8), 26−28.

Saha, B.C., 2003. Hemicellulose bioconversion. J. Ind. Microbiol. Biotechnol. 30, 279−291.

Saska, M., Ozer, E., 1995. Aqueous extraction of sugarcane bagasse hemicelluloses and production of xylose syrup. Biotechnol. Bioeng. 45, 517−523.

Schlesinger, R., Götzinger, G., Sixta, H., Friedl, A., Harasek, M., 2006. Evaluation of alkali resistant nanofiltration membranes for the separation of hemicelluloses from concentrated alkaline process liquors. Desalination 192, 303−314.

Schönberg, C., Oksanen, T., Suurnäkki, A., Kettunen, H., Buchert, J., 2001. The importance of xylan for the strength properties of spruce kraft pulp fibers. Holzforschung 55 (6), 639−644.

Scott, R.W., 1989. Influence of cations and borate on the alkali extraction of xylan and glucomannan from pine pulps. J. Appl. Polym. Sci. 38 (5), 907−914.

Shimizu, K., Sudo, K., Ono, H., Ishihara, M., Fujii, T., Hishiyama, S., 1998. Integrated process for total utilization of wood components by steam explosion pretreatment. Biomass Bioenergy 14 (3), 195–203.

Sinquefield, S., 2005. *In situ* causticizing for black liquor gasification. Phase 2 Tropical Report, February 1, 2004–October 31, 2005.

Sricharoenchaikul, V., 2001. Fate of Carbon-Containing Compounds from Gasification of Kraft Black Liquor with Subsequent Catalytic Conditioning of Condensable Organics. Georgia Institute of Technology, Atlanta GA, USA, PhD Dissertation.

Stigsson, L., 1998. Chemrec black liquor gasification. International Chemical Recovery Conference. 1–4 June, Tampa, FL.

Suresh, P., 2002. Biomass Gasification for Hydrogen Production—Process Description and RESEARCH Needs. IEA Thermal Gasification Task Leader Gas Technology Institute, IL.

Sun, R.C., Fang, J.M., Tomkinson, J., Geng, Z.C., Liu, J.C., 2001. Fractional isolation, physicochemical characterization and homogeneous esterification of hemicelluloses from fast-growing poplar wood. Carbohyd. Polym. 44, 29–39.

Swatloski, R.P., Spear, S.K., Holbrey, J.D., Rogers, R.D., 2002. Dissolution of cellulose with ionic liquids. J. Am. Chem. Soc. 124 (18), 4974–4975.

Tampier, M., Smith, D., Bibeau, E., Beauchemin, P.A., 2004. Identifying Environmentally Preferable Uses for Biomass Resources—Stage 1 Report: Identification of Feedstock-to-Product Threads. Envirochem Services Inc., North Vancouver, BC, Report.

Thorp, B., 2005a. Transition of mills to biorefinery model creates new profit streams. Pulp Pap. November, 35–39.

Thorp, B., 2005b. Biorefinery offers industry leaders business model for major change. Pulp Pap. 79 (11), 35–39.

Thorp, B., Raymond, D., 2005. Forest biorefinery could open door to bright future for P&P industry. PaperAge 120 (7), 16–18.

Thorp, B.A., Thorp, B.A., Murdock-Thorp, L.D., 2008. A compelling case for integrated biorefineries. <http://www.epoverviews.com/oca/Compellingcaseforbiorefineries.pdf>. (accessed 15.11.2012).

Thurso Project, 2010. Forthress specialty cellulose. Available from: <http://specialtycellulose.com>. (accessed 10.9.2012).

Tomani, P., 2010. Lignin extraction from black liquor. Forty-Third Pulp and Paper International Congress and Exhibition, 2010 (ABTCP-TAPPI 2010). 4–6 October, Sao Paulo, Brazil.

Tucker, M.P., Kim, K.H., Newman, M.M., Nguyen, Q.A., 2003. Effects of temperature and moisture on dilute-acid steam explosion pretreatment of corn stover and cellulase enzyme digestibility. Appl. Biochem. Biotechnol. 105, 165–177.

Tucker, P., 2002. Changing the balance of power. Solutions 85 (2), 34–38.

Vakkilainen, E.K., Kankkonen, S., Suutela, J., 2008. Advanced efficiency options: increasing electricity generating potential from pulp mills. Pulp Pap. Can. 109 (4), 14–18.

van Heiningen A., 2005. Hemicellulose extraction and its integration in pulp production. Forest Products Industry of the Future, Quarterly Status Reports. September 30, US Department of Energy website: <http://www.eere.energy.gov/industry/forest/pdfs/quarterlyhighlights.pdf>.

van Heiningen, A., 2006. Converting a kraft pulp mill into an integrated biorefinery. Pulp Pap. Can. 107 (6), T141–T146.

Wai, C.M., Gopalan, A.S., Jacobs, H.K., 2003. An introduction to separations and processes using supercritical carbon dioxide. In: ACS Symposium Series, 860 (Supercritical Carbon Dioxide), 2–8, American Chemistry Society.

Wallmo, H., Theliander, H., 2007. The LignoBoost process: comments on key-operations. International Chemical Recovery Conference: Efficiency and Energy Management. 29 May–1 June, Quebec City, QC, Canada. pp. 333–335.

Warnqvist, B., Delin, L., Theliander, H., Nohlgren, I., 2000. Teknisk Ekonomisk Utvärdering Avsvartlutförgasningsprocesser. Värmeforsk service AB, Stockholm.

Warnqvist, B., Delin, L., Theliander, H., Nohlgren, I., 2001. Techno-economical evaluation of black liquor gasification processes. International Chemical Recovery Conference, Whistler, BC, Canada.

Weil, J.R., Sarikaya, A., Rau, S.L., Goetz, J., Ladisch, C.M., Brewer, M., et al., 1997. Pretreatment of yellow poplar sawdust by pressure cooking in water. Appl. Biochem. Biotechnol. 68 (1/2), 21–40.

Weil, J.R., Brewer, M., Hendrickson, R., Sarikaya, A., Ladisch, M.R., 1998. Continuous pH monitoring during pretreatment of yellow poplar wood sawdust by pressure cooking in water. Appl. Biochem. Biotechnol. 70–72, 99–111.

Whitty, K., 2005. Black liquor gasification: development and commercialization update. ACERC Annual Conference. Provo Utah.

Whitty, K., Baxter, L., 2001. State of the art in black liquor gasification technology. Joint International Combustion Symposium. 9–12 September, Kauai, Hawaii.

Whitty, K., Nilsson, A., 2001. Experience from a high temperature, pressurized black liquor gasification pilot plant. International Chemical Recovery Conference. 11–14 June, Whistler, Canada.

Whitty, K., Verrill, C.L., 2004. A historical look at the development of alternative black liquor recovery technologies and the evolution of black liquor gasifier designs. International Chemical Recovery Conference. 6–10 June, Charleston, SC.

Wising, U., Stuart, P.R., 2006. Identifying the Canadian forest biorefinery. Pulp Pap. Can. 107 (6), 25–30.

Wyatt, V.T., Bush, D., Lu, J., Hallett, J.P., Liotta, C.L., Eckert, C.A., 2005. Determination of solvatochromic solubility parameters for the characterization of gas-expanded liquids. J. Supercrit. Fluids 36 (1), 16–22.

Yan, J., Dahlquist, E., Jin, H., Gao, L., Tu, S., 2007. Integration of large scale pulp and paper mills with CO_2 mitigation technologies. The Third International Green Energy Conference, Västerås, Sweden.

Yanagisawa, M., Shibata, I., Isogai, A., 2005. SEC-MALLS analysis of softwood kraft pulp using LiCl/1,3-dimethyl-2-imidazolidinone as an eluent. Cellulose 12 (2), 151–158.

Yunqiao, P., Zhang, D., Singh, P.M., Ragauskas, A.J., 2008. Biofuels Bioprod. Bioeng. 2 (1), 58–73.

Zeng, L., van Heiningen, A., 1996. Proceedings of Eighty-Second Annual Conference, Tech. Sect.,CPPA, Montreal, 1996, A259.

Zou, X., Avedisian, M., van Heiningen, A., 1992. Kinetics of the direct causticization reaction between sodium carbonate and titanium dioxide. AIChE Forest Products Symposium Series, AIChE, New York, NY.

CHAPTER 4

Products from Hemicelluloses*

4.1 ETHANOL

For conversion to ethanol, hemicelluloses are first converted to sugars by using acid hydrolysis and enzymatic hydrolysis processes. In these processes, monosaccharides are produced that are converted to ethanol using fermentation (Kim, 2005; Nguyen et al., 2000; Wright and Power, 1987; Wyman and Goodman, 1993). Depending on what technologies are optimized for the pre-extraction of hemicelluloses from wood chips, an acid hydrolysis of polysaccharides to hexoses and pentoses may be preferred. The pentose sugar xylose is the major carbohydrate component of hemicellulose in a wide variety of lignocellulosic biomass species. Consequently, the ability to ferment xylose is an important characteristic

---

*Some excerpts are taken from Bajpai (2012). Biotechnology for Pulp and Paper Processing with kind permission from Springer Science + Business Media

Biorefinery in the Pulp and Paper Industry. DOI: http://dx.doi.org/10.1016/B978-0-12-409508-3.00004-3

of microorganisms, being considered for use in large-scale fermentation-based hemicellulose conversion processes.

D-xylose is not readily utilized as D-glucose for the production of ethanol by microorganisms such as *Saccharomyces cerevisiae*. Bacteria, yeast, and mould differ in their mode of conversion of D-xylose to xylulose, the initial step in xylose fermentation. Few yeasts use the enzymes xylose reductase to reduce xylose to xylitol which is subsequently oxidized to D-xylulose by xylitol dehydrogenase. Yeasts such as *Pichia*, *Kluyveromyces*, and *Pachysolen* are reported to ferment xylose to ethanol under micro-aerophilic condition. The general requirements of an organism to be used in ethanol production is that it should give a high ethanol yield, a high productivity, and be able to withstand high ethanol concentrations in order to keep distillation costs low. In addition to these general requirements, inhibitor tolerance, temperature tolerance, and the ability to utilize multiple sugars are also essential.

The enzymatic hydrolysis of pretreated cellulosic biomass has been commercialized for the processing of wheat straw to bioethanol and is being actively pursued for other agricultural waste resources (Tolan, 2003). An important consideration for hemicellulose pre extraction and depolymerization treatment protocol is to reduce byproducts that are inhibitors of the fermentation of sugars to ethanol, such as furans, carboxylic acids, and phenolic compounds (Palmqvist and Hahn-Hägerdal, 2000). Some inhibitors are present in the raw material, but others can be formed during the hydrolysis process (Klinke et al., 2004). The nature, composition, and concentration of these compounds are dependent on the hydrolysis conditions and may have a profound influence on the fermentation production rate of biofuels from the hydrolyzate (Taherzadeh et al., 2000a,b). There are several strategies: chemical, physical, or biological for dealing with the inhibitors in hydrolyzates (Alriksson et al., 2005; Horváth et al., 2005; Larsson et al., 1999; Persson et al., 2002). Methods commonly used for detoxification of hydrolyzates before fermentation are as follows (Canilha et al., 2004; Cantarella et al., 2004; Chandel et al., 2007; Cruz et al., 1999; Gonzalez et al., 2004; Gutiérrez et al., 2006; Jönsson et al., 1998; Palmqvist and Hahn- Hägerdal, 2000; Parajo et al., 1996; Villarreal et al., 2006; Wilson et al., 1989):

• *Evaporation*
  Evaporation removes acetic acid, furfural, and other volatile components in the hydrolyzates. Converti et al. (2000) hydrolyzed the

Eucalyptus globulus wood using steam explosion and dilute acid treatment techniques at 100°C, followed by boiling or evaporating the obtained hydrolyzate for 160 min. The concentration of acetic acid and furfural reduced from 31.2 to 1.0 g/l and from 1.2 to 0.5 g/l, respectively. These are the inhibitory levels for the fermentation of xylose to xylitol by *Pachysolen tannophilus* strain, showing that in this case the simple evaporation method is sufficient to eliminate the inhibition of acetic acid and furfural.

- *Solvent extraction*
  Solvent extraction with ethyl acetate was found to be effective for removal of all of the inhibitory compounds except for the residual acetic acid (Wilson et al., 1989). Palmqvist and Hahn- Hägerdal (2000) used ethyl acetate extraction for removing 56% acetic acid and all of furfural, vanillin, and 4-hydroxybenzoic acid.

- *Overliming with calcium hydroxide*
  In the overliming process, Palmqvist et al. (2000) and Converti et al. (2000) detoxified the hydrolyzate by addition of $Ca(OH)_2$ to adjust the pH to $9-10$. This resulted in precipitation of inhibitory compounds. After filtration, the pH of the resulting hydrolyzate is readjusted to 5.5 with dilute H_2SO_4 ready for fermentation. This process is found to be very effective, but it has a drawback in that it produces a large amount of gypsum (Aden et al., 2002).

- *Activated charcoal*
  Converti et al. (2000) used the activated charcoal adsorption process to remove the inhibitory compounds. Lignin-derived inhibitory compounds are adsorbed on activated charcoal, and after its adsorption saturation, the charcoal is reactivated or regenerated by heating, for example, by boiling it in distilled water for 3 h. Parajo et al. (1996) studied the effect of different operating variables—hydrolyzate concentration, adsorbent charge (hydrolyzate:charcoal ratio), and adsorption time—on the subsequent fermentation of xylose to xylitol by the yeast *Debaryomyces hansenii*. It was found that a hydrolyzate:charcoal ratio of 205 g/g was sufficient for improving subsequent fermentation.

- *Ion-exchange resins*
  Ion-exchange resin (IER) is a well-known method for removal of inhibitory compounds. The elimination of acetic acid inhibition of D-xylose fermentation by *Pichia stipitis* with IER was studied by Van Zyi et al. (1991). The inhibition degree was found to be dependent on the acetic acid concentration, the oxygen availability, and the

pH value. A comparison was made between the fermentation of an untreated acid hydrolyzate of sugarcane bagasse and the fermentation of the hydrolyzate treated by an anion-exchange resin with the removal of 84% of the acetic acid. It was found that the former ethanol yield was 0.27 g/g sugar, while the latter ethanol yield increased by 0.36 g/g sugar.

- *Enzymatic detoxification*
  Enzymatic treatment is also found to be effective for removal of phenolic compounds, Jönsson et al. (1998) used the laccase enzyme for removing phenolics from Willow hydrolyzates treated with steam and SO_2 (Jonsson et al., 1998). Laccase treatment can remove most of the phenolics, but not acetic acid, furfural, and hydroxyl methyl furfural (HMF). For example, around 80% of the phenolic compounds was removed from sugarcane bagasse hydrolyzates obtained by steam explosion with the phenoloxidase laccase (Martin et al., 2002). In addition, reductive detoxification of furfural to the less toxic furfuryl alcohol can be performed by using the ethanologenic bacterium *Escherichia coli* strain LYO1 (Gutiérrez et al., 2006).

Several researchers have made comparisons between different detoxification methods for selecting an efficient detoxification approach.

Chandel et al. (2007) studied the detoxification of sugarcane bagasse hydrolyzate to improve ethanol production by *Candida shehatae* NCIM 3501. Five detoxification methods including neutralization, overliming, activated charcoal, IER, and enzymatic detoxification using laccase were compared. They found that ion-exchange treatment was most effective in removing furans (63.4%), total phenolics (75.8%), and acetic acid (85.2%). Activated charcoal removed 38.7%, 57.0%, and 46.8% of furans, phenolics, and acetic acid, respectively. Laccase treatment removed 77.5% of total phenolics, but it could not reduce the furans and acetic acid contents in its hydrolyzates. Overliming reduced 45.8% furans and 35.8% phenolics in its treated hydrolyzate, but not acetic acid. Fermentation of hydrolyzates detoxified by different methods with *Candida shehatae* NCIM 3501 showed that the ethanol yields obtained by different detoxification treatments were in the following decreasing order: ion exchange (0.48 g/g) > activated charcoal (0.42 g/g) > laccase (0.37 g/g) > overliming (0.30 g/g) > neutralization (0.22 g/g).

Cantarella et al. (2004) made a comparison between $Ca(OH)_2$ overliming, water rinsing, water−ethyl acetate two-phase contacting, and *in situ* detoxification procedures with high-level yeast inocula, to eliminate the inhibition problem in saccharification of cellulose from steam-exploded (SE) poplar wood to glucose by enzyme cellulases and fermentation of glucose to ethanol by *S. cerevisiae* (Baker's yeast). The water-rinsing treatment removed water-soluble inhibitors, thus enhanced the enzymatic hydrolysis of SE substrate. Obviously, however, this method is suitable for washing away the inhibitors in cellulose hydrolyzates, but not for removal of inhibitors in hemicellulosic prehydrolyzates as most pentose sugars are also water soluble. This comparison showed that overliming with $Ca(OH)_2$ is the most efficient method among those investigated.

Villarreal et al. (2006) investigated detoxification of Eucalyptus hemicellulose hydrolyzate, that is, removal of acetic acid, furfural, HMF, and phenolics, for xylitol production by *Candida guilliermondii* with active charcoal and a series of IER columns composed of four different resins (alternate cationic and anionic) in sequence. They found that IER can remove all inhibitory components (aliphatic acids, furan derivatives, and phenols) without significant loss of sugar, superior to activated charcoal.

In situ detoxification by using extractive-fermentation, membrane pervaporation-bioreactor hybrid, and vacuum membrane distillation (VMD)-bioreactor hybrid processes can effectively separate ethanol and remove inhibitors simultaneously. In conclusion, evaporation is a simple way to remove acetic acid, furfural, and other volatile components, but it is difficult to remove the heavier components with higher boiling points. Extraction with solvent is efficient in removing all the inhibitors, but it needs additional solvent and solvent recovery for recycle use. Activated charcoal adsorption can remove the phenolic compounds, but it is not so efficient in removal of acetic acid and furfural. Overliming and IER are more effective procedures for removal of different inhibitors from hydrolyzates, but the overliming process leads to production of large amount of gypsum. The IER method is currently the best method for detoxification because of its high detoxification efficiency, easy operation, and flexible combination of different anion and cation exchangers, while the enzymatic treatment can possibly be the future choice. In addition, extractive-fermentation, membrane pervaporation-bioreactor hybrid, and VMD-bioreactor hybrid are very promising approaches to remove inhibitory compounds in addition to increasing ethanol yield.

The hydrolyzed hemicellulose sugar solution will finally need to undergo fermentation for the production of ethanol. The microorganisms that are able to ferment sugars to ethanol can be either yeasts or bacteria (Kuyper et al., 2005a,b; Senthilkumar and Gunasekaran, 2005). Recent advances in genetic engineering, forced evolution, and mutation and selection strategies have enhanced the biological utilization of hexoses and pentoses for the biological production of ethanol. The well-documented fermentation of wood hydrolyzates to ethanol provides a strong technical basis from which practical fermentation technologies can be designed for the conversion of pre-extracted wood hemicelluloses to ethanol. The fermentation of dilute acid hydrolyzates from aspen, birch, willow, pine, and spruce using *S. cerevisiae* has been reported (Taherzadeh et al., 1997). These wood hydrolyzates contained varying amounts of xylose, glucose, and mannose, and the efficiency of fermentation varied substantially, depending upon wood species employed. The use of other yeast and fungi for the production of ethanol from wood hydrolyzates has also been reported, and their efficiencies and cost-performance properties continue to be enhanced (Brandberg et al., 2004; Millati et al., 2005; Sreenath and Jeffries, 1999; Zaldivar et al., 2001).

The concept of hemicellulose pre extraction prior to pulping has been funded by a consortium of large pulp and paper manufacturers and is being operated under the auspices of Agenda 2020. In the USA, wood chip pre extraction technologies could make available to the biofuels industry about 14 million tons of hemicelluloses annually while at the same time enhancing the production of kraft pulps (Ragauskas et al., 2006). These extractable hemicelluloses could provide a valuable, high-volume resource of sugars for bioethanol production generating approximately 20–40 million gallons ethanol per year per mill (Amidon et al., 2007). Thorp (2005) and Thorp and Raymond (2005) have reported that the potential annual production of ethanol from pre extraction of hemicellulose could approach 2 billion gallons of ethanol per year. Extracting the hemicellulose from the wood chips prior to pulping and depositing the oligomer portion onto the pulp stream after the digester could increase pulp yield by 2%, resulting in approximately $600 million a year in extra pulp production (US Department of Energy, 2006).

Research studies have already established the viability of extracting hemicelluloses from wood chips prior to kraft pulping for dissolving pulps. The challenge for the biofuels and forest products industries is to develop optimized pre-extraction technologies that provide a hemicellulose stream for biofuels production and a lignocellulosics stream for pulp production. This vision will, undoubtedly, require a cooperative research program with multipartner stakeholders. These efforts have already begun and will accelerate in the near future, given the significant benefits to all interested parties (Bajpai, 2012).

4.2 FURFURAL

Furfural represents a renewable building block chemical that is currently regaining attention as a biobased alternative for the production of industrial and household chemicals (Christopher, 2012; Mamman et al., 2008; Marinova et al., 2010; Menon et al., 2010). Furfural is produced by acid hydrolysis of pentosans. It is the only organic compound derived from biomass that can replace the crude oil-based organics used in industry. Furfural is used as a chemical feedstock for furfuryl alcohol (production of furan resins), in the manufacture of furan (as an intermediate in the synthesis of pharmaceuticals, agricultural chemicals, stabilizers, and fine chemicals), as well as a solvent for refining lubricating oils, as a decolorizing and wetting agent. One of the commercially available methods to generate furfural is the Westpro modified Huaxia technology. The raw material, that could be either forest or agricultural, is charged to steel reactors or digesters. The furfural thus formed is entrained with steam and the furfural-saturated steam is condensed. The condensed solution is then fed to a furfural azeotropic distillation column, where the condensate is separated into two fractions. The light water phase is refluxed and the heavy furfural phase undergoes refining by continuous azeotropic distillation. The maximum furfural yield that could be obtained from wood using the Westpro modified Huaxia technology is 8%. This process also generates byproducts, such as acetic acid, acetone, and methyl alcohol, which are not considered in this study. China is the biggest supplier of furfural, and accounts for around half of the global capacity. The world production of furfural in 2005 was about 250,000 t/a, at a stable price of $1,000/t (Win, 2005).

4.3 XYLITOL

Xylitol is a five-carbon sugar alcohol. It occurs naturally in fruits, in low amounts.

It has attracted much attention because of its application in food and confectionery, in the production of oral hygiene products (mouthwash and toothpaste), pharmaceuticals, dietetic products, and cosmetics (Saha and Bothast, 1997). The bulk of xylitol produced is consumed in various food products such as chewing gum, candy, soft drinks, and ice cream. It gives a pleasant, cool, and fresh sensation due to its high negative heat of solution. It is currently produced by chemical reduction in alkaline conditions of the xylose derived mainly from wood hydrolyzate (Malaja and Hamalainen, 1977). The recovery of xylitol from the xylem fraction is about 50−60% or 8−15% of the raw material employed (Winkelhausen and Kuzmanova, 1998). The value depends on the xylem content of the raw material.

Xylitol is produced chemically by hydrogenation of xylose, which converts the sugar aldehyde into a primary alcohol (Karimkulova et al., 1989). Hydrogenation is carried out at high pressures (up to 50 atm), high temperature (80−140°C) using expensive catalysts (Raney nickel) and expensive purification processes (Mikkola et al., 2000). The xylitol yields are low, on average 50−60% from xylan. Drawbacks of the chemical process are the requirements of high pressure (up to 50 atm) and temperature (80−140°C), use of an expensive catalyst (Raney nickel), and use of extensive separation and purification steps to remove the byproducts, which are derived mainly from the hemicellulose hydrolyzate (Meinander et al., 1994). The drawbacks of the chemical process can be overcome by using a biological route of xylitol production that is carried out by microorganisms at low temperature (30−35°C). The microbial conversion employs naturally fermenting yeasts (*Candida*) such as *C. tropicalis and C. guilliermondii* with yield of 65−90% from xylan. Alternatively, recombinant strains containing a xylose reductase gene (i.e., recombinant *S. cerevisiae*) can be used with a very high production yield of 95% from the theoretical maximum.

The microbial production of xylitol has been reviewed by Saha (2003). Some natural xylose-fermenting yeasts known to produce xylitol are *C. boidini, C. guilliermondii, C. tropicalis, C. parapsilosis*, and *D. hansenii* (Saha and Bothast, 1997; Silva et al., 1998). Many yeasts and mycelial fungi possess NADPH-dependent xylose reductase, which catalyzes the reduction of xylose to xylitol as a first step in xylose metabolism (Chiang and Knight, 1960). Xylitol can be subsequently oxidized to xylulose by the action of xylitol dehydrogenase, which preferentially uses NAD as an acceptor (Hofer et al., 1971). In xylose-fermenting yeasts, the initial reactions of xylose metabolism appear to be rate limiting (Nolleau et al., 1995). This results in accumulation of xylitol in the culture medium. The degree varies with the culture conditions and the yeast strain used (Van Dijken and Scheffers, 1986). A surplus of NADH during transient oxygen limitation inhibits the activity of xylitol dehydrogenase resulting in xylitol accumulation (Granstorm et al., 2001).

The fermentation of sugarcane bagasse hemicellulose hydrolyzate to xylitol by a hydrolyzate-acclimatized yeast strain *Candida* sp. B-22 was studied by Chen and Gong (1985). With this strain, a final xylitol concentration of 94.74 g/l was obtained from 105.35 g/l xylose in hemicel-luloses hydrolyzate after 96 h of incubation. *Candida guilliermondii* FTI 20037 was able to ferment a sugarcane bagasse hydrolyzate pro-ducing 18.4 g/l xylitol from 29.5 g/l of xylose, at a production rate of 0.38 g/l/h (Pfeifer et al., 1996). This lower value, compared to that (0.66 g/l/h) of the synthetic medium, may be attributed to the various toxic substances that interfere with microbial metabolism (e.g., acetic acid). Researchers studied (Dominguez et al., 1996) different treatments (neutralization, activated charcoal and neutralization, cation-exchange resins, and neutralization) of sugarcane bagasse hemicelluloses hydroly-zate to overcome the inhibitory effect on xylitol production by *Candida* sp. 11-2. The highest xylitol productivity (0.205 g/l/h), corresponding to 10.54 g/l, was obtained from hydrolyzates treated with activated charcoal (initial xylose, 42.96 g/l). To obtain higher xylitol productivity, treated hydrolyzates were concentrated by vacuum evaporation in rotavapor to provide higher initial xylose concentration. The rate of xylitol production increased with increasing initial xylose concentration from 30 to 50 g/l reaching a maximum of 28.9 g/l after 48 h fermentation. The decrease in xylitol production was dramatic with further increases in the initial xylose concentration. Parajo et al. (1997) later reported a xylitol production of 39–41 g/l from concentrated *Eucalyptus globulus* wood acid hydrolyzate containing 58–78 g xylose/l by *D. hansenii* NRRL Y-7426 using an initial cell concentration of 50–80 g/l.

Carvalho et al. (2002) obtained maximum xylitol concentration of 20.6 g/l with a volumetric productivity of 0.43 g/l/h and yield of 0.47 g/g after 48 h fermentation during batch xylitol production from con-centrated sugarcane bagasse hydrolyzate and C. *guilliermondii* cells,

Table 4.1 Xylitol Production from Detoxified Hemicellulosic Hydrolyzates by Fermentation

Substrate	Yeast	Fermentation Time (h)	Xylose (g/l)	Xylitol (g/l)	Xylitol (g/g)
Wood	Debaryomyces hansenii NRRL Y-7426	78	78	41	0.73
Hardwood	Pachysolen tannophillus	96	89	39.5	
Rice straw	Candida guilliermondii FTI 20037	72	64	37.6	0.62
Sugarcane bagasse	Candida sp. B-22	96	105.4	96.8	0.89
Corncob	Candida parapsilosis	59	50	36	0.72

Source: Based on Saha and Bothast (1997).

immobilized in calcium-alginate beads. The production of xylitol from various detoxified hemicellulosic hydrolyzates is shown in Table 4.1. A number of recombinant *S. cerevisiae* strains have been created by expressing the xylose reductase gene (*XTLJ*) from *P. stipitis* and *C. shehatae* and production of xylitol from xylose by these recombinant strains in batch and fed-batch fermentations have been investigated (Chung et al., 2002; Govinden et al., 2001; Hallborn et al., 1991, 1994; Lee et al., 2000). These strains converted xylose to xylitol with over 95% yield. These recombinant strains in batch and fed-batch fermentations have been investigated. These strains converted xylose to xylitol with over 95% yield.

Xylitol recovery from fermented sugarcane bagasse hydrolyzate was studied by Gurgel et al. (1995). The best clarifying treatment was found by adding 20 g activated carbon to 100 ml fermented broth at 80°C for 1 h at pH 6.0. The clarified medium was treated with IERs after which xylitol crystallization was attempted. The IERs were not efficient but the crystallization technique showed good performance, although the crystals were involved in a viscous and colored solution. Another research group reported (Faveri et al., 2002a) xylitol recovery by crystallization from synthetic solutions and fermented hemicellulose hydrolyzates. They concluded that xylitol separation by crystallization from fermented hemicellulose hydrolyzate is feasible.

The method involves evaporation of dilute solution up to supersaturation, cooling of the supersaturated solution, separation of crystals by centrifuge, and final filtration. Using two sets of tests on xylitol–xylose synthetic solutions and one set on fermented hardwood hemicelluloses hydrolyzate, the best results in terms of either crystallization yield (0.56) or purity degree (1.00) were obtained with quite concentrated solutions of 730 g/l at a relatively high temperature ($-5°C$).

The microbially-produced xylitol requires less purification than the chemical process (Prakasham et al., 2009). Due to its anticariogenic and antiplaque action (Trahan, 1995), xylitol is used around the world as a sweetener in chewing gums and oral hygiene products such as toothpaste, fluoride tablets, and mouthwashes. More than 10% of its use in sugar-free chewing gums which have a world market of more than $12 million per annum. Due to its structure, xylitol is a nonfermentable sugar alcohol with dental health benefits in caries prevention, showing superior performance to other polyols (polyalcohols). Its plaque-reducing effect is manifested by attracting and starving harmful microorganisms because cariogenic bacteria prefer fermentable six-carbon sugars as opposed to the nonfermentable xylitol (Milgrom et al., 2006).

Possessing approximately 40% less food energy, xylitol is a low-calorie alternative to table sugar. Absorbed more slowly than sugar, it does not contribute to high blood sugar levels or the resulting hyperglycemia caused by insufficient insulin response. Its glycemic index (GI) is approximately 10-fold lower than that of sucrose. This characteristic has also proven beneficial for people suffering from metabolic syndrome, a common disorder that includes insulin resistance, hypertension, hypercholesterolemia, and an increased risk for blood clots. Xylitol also has potential as a treatment for osteoporosis—it prevents weakening of bones and improves bone density (Mattila et al., 2002). Studies have shown xylitol chewing gum can help prevent ear infections (Uhari et al., 1998). When bacteria enter the body, they adhere to the tissues using a variety of sugar complexes. The open nature of xylitol and its ability to form many different sugarlike structures appears to interfere with the ability of many bacteria to adhere. Xylitol is also one of the building block chemicals that can be used in production of ethylene glycol, propylene glycol, lactic acid, and for synthesis of unsaturated polyester resins, antifreeze, etc.

4.4 2,3-BUTANEDIOL

2,3-Butanediol, otherwise known as 2,3-butylene glycol (2,3-BD), is a valuable chemical feedstock because of its application as a solvent, liquid fuel, and as a precursor of many synthetic polymers and resins. With a heating value of 27,200 J/g, 2,3-BD compares favorably with ethanol (29,100 J/g) and methanol (22,100 J/g) for use as a liquid fuel and fuel additive (Tran and Chambers, 1987). Dehydration of 2,3-BD yields the industrial solvent methyl ethyl ketone, which is much more suited as a fuel because of its much lower boiling point. Further dehydration yields 1,3-butanediene, which is the starting material for synthetic rubber and is also an important monomer in the polymer industry (Maddox, 1996). During the Second World War, it was needed for conversion to 1,3-butanediene. Methyl ethyl ketone can be hydrogenated to yield high octane isomers suitable for high-quality aviation fuels. Diacetyl, formed by catalytic dehydrogenation of the diol, is a highly valued food additive (Magee and Kosaric, 1987). A wide variety of chemicals can also be easily prepared from 2,3-BD (Gong et al., 1997; Yu and Saddler, 1985). There is interest in industrial-scale production of 2,3-BD from various agricultural residues as well as from logging, pulp and paper and food industry wastes (Magee and Kosaric, 1987).

2,3-BD can occur in two enantiomeric forms: D-(−) and L-(+) as well as an optically inactive mesoform. *Bacillus polymyxa* produces D-(−)2,3-BD whereas *Klebsiella pneumoniae* (*Aerobacter aerogenes*) produce mesoform and also some of the L-(+) form. *Bacillus subtilis*, *Seratia marcescens*, and *Aerobacter hydrophila* produce mixtures of different forms (Kosaric and Velikonja, 1995). Saha and Bothast (1999) isolated *Enterobacter cloacae* NRRL B-23289 from local decaying wood/corn soil samples while screening for microorganisms for conversion of arabinose to fuel ethanol. The major product of fermentation by the bacterium was meso-2,3-BD. In a typical fermentation, a 2,3-BD yield of 0.43 g/g arabinose was obtained at an initial arabinose concentration of 50 g/l. The bacterium utilized sugars from acid plus enzyme saccharified corn fiber and produced 2,3-BD (0.35 g/g available sugars). It also produced 2,3-BD from dilute acid pretreated corn fiber acid by simultaneous saccharification and fermentation (0.34 g/g theoretical sugars). The 2,3-BD yield (0.35−0.43 g/g sugar) by the newly isolated *Enterobacter cloacae* NRRL B-23289 compares favorably with other 2,3-BD-producing organisms reported in the literature (0.30−0.45 g/g sugar) (Maddox, 1996).

Butanediol is produced during oxygen-limited growth, by a fermentative pathway known as the mixed acid-butanediol pathway. The 2,3-BD pathway and the relative proportions of acetoin and butanediol serve to maintain the intracellular NAD/NADH balance under changing culture conditions. The theoretical maximum yield of 2,3-BD from monosaccharides is 0.5 g/g (Jansen et al., 1984). The efficient biological conversion of all available sugars in agricultural biomass residues to fuels and chemicals is crucial to the efficiency of any process intended to compete economically with petrochemical products (Yu and Saddler, 1985). The high boiling point of 2,3-BD, its high affinity for water, and the dissolved and solid substances of the fermentation broth make it difficult for 2,3-BD to be purified and recovered from fermentation slurry (Syu, 2001) Various methods, such as solvent extraction, liquid–liquid extraction, and salting out, have been used to recover butanediol. Another feasible method to recover butanediol is countercurrent stream stripping (Garg and Jain, 1995).

4.5 ORGANIC ACIDS

In the last decade, microbially-produced organic acids (Mattey, 1992) have found increased use in the food industry and as raw materials for manufacture of biodegradable polymers (Magnuson and Lasure, 2004). For instance, the production of D-lactic acid as well as L-lactic acid is of significant importance for the practical application of poly-lactic acid, which is an important raw material for bioplastics (Okano et al., 2009). Organic acids are used in food preservation because of their effects on bacteria (Dibner and Butin, 2002). The nondissociated (nonionized) organic acids can penetrate the bacteria cell wall and disrupt the normal physiology of certain types of bacteria such as *E. coli*, *Salmonella*, and *Campylobacter* species that are pH-sensitive and cannot tolerate a wide internal and external pH gradient. Upon passive diffusion of organic acids into the bacteria where the pH is near or above neutrality, the acids dissociate and the cations lower the bacteria internal pH, leading to situations that impair or stop the growth of bacteria. Furthermore, the anions of the dissociated organic acids accumulate within the bacteria and disrupt their metabolic functions leading to osmotic pressure increase that is incompatible with the bacterial survival. For example, lactic acid and its sodium and potassium salts are widely used as antimicrobials in food products, in particular, meat and poultry such as ham and sausages. Tables 4.2 and 4.3 summarize the major production organisms, substrates, and uses of organic acids.

Table 4.2 Organic Acids Produced from Xylose

Organic Acid	Microorganism	Substrate	State
Lactic acid	L. delbrueckii	Xylose, glucose, starch, cellulose, newspaper, MSW (xylose, mannose)	Commercial
	A. oryzae		
Citric acid	A. niger	Sugarcane molasses, corn syrups, lignocellulose, agri- and food waste	Commercial
Gluconic acid	A. niger	Glucose, glucose corn syrups	Commercial
Itaconic acid	A. terreus	Sugarcane molasses, corn syrups, xylose	Commercial
Aspartic acid	E. coli	Fumaric acid + NH_3	Commercial
Malic acid	Brevibacterium	Fumaric acid	Commercial
Succinic acid	A. succiniciproducens	Glucose, sugarcane molasses	Experimental
	A. succinogenes, E. coli (recombinant)	Glucose, xylose	
Fumaric	Rhyzopus spp.	Glucose, sucrose, sugarcane molasses, corn syrups, starch, xylose	Experimental

Source: Based on Christopher (2012).

Table 4.3 Application of Organic Acids

Lactic acid
Acidulant, flavor enhancer, food preservative, feedstock for calcium stearoyl-2-lactylates (baking), ethyl lactate (biodegradable solvent), and polylactic acid plastics (100% biodegradable) for packaging, consumer goods, biopolymers (approved by FDA). Estimated US consumption 30 million pounds with 6% growth pa. Potential demand 5.5 billion pounds as very large volume—commodity chemical

Citric acid
70% in food, confectionary, and beverage products, 30% pharmaceuticals (anticoagulant blood preservative, antioxidant) and metal cleaning. Selling price decreased with market shift from pharmaceuticals to food applications (879,000 MT produced in 2002)

Aspartic acid
For synthesis of aspartame, monomer for manufacture of polyesters, polyamides, polyaspartic acid as a substitute for EDTA with potential market of $450 million per year

Itaconic acid
Feedstock for syntheses of polymers for use in carpet backing and paper coating N-substituted pyrrolidinones for use in detergents and shampoos. Cements comprising copolymers of acrylic and itaconic acid

Fumaric acid
For manufacture of synthetic resins, biodegradable polymers, intermediate in chemical and biological synthesis

Malic acid
Acidulant in food products, citric acid replacement, raw material for manufacture of biodegradable polymers, for treatment of hyperammonemia, liver dysfunction, component for amino acid infusions

Succinic acid
Acidulant, pH modifier, flavoring and antimicrobial agent, ion chelator in electroplating to prevent metal corrosion, surfactant, detergent, foaming agent, for production of antibiotics, amino acids, and pharmaceuticals. Market potential of 270,000 t in 2004, US domestic market estimated at $1.3 billion per year with 6—10% annual growth

Source: Based on Christopher (2012).

Lactic acid is used in the food, pharmaceutical, and cosmetic industries. It is a component of biodegradable plastic polylactate, the market for which is expected to grow significantly. It is also used as a source of lactic acid polymers that are being used as biodegradable plastics (Brown, 2006; Datta et al., 1995). The physical properties and stability of polylactides can be controlled by adjusting the proportions of the L(+)- and D(−)-lactides (Tsuji, 2002). Optically-pure lactic acid is currently produced by the fermentation of glucose derived from corn starch using various lactic acid bacteria (Carr et al., 2002; Hofvendahl and Hahn-Hägerdal, 2000). However, the fastidious lactic acid bacteria have complex nutritional requirements (Chopin, 1993) and the use of corn is not favored as the feedstock competes directly with the food and feed uses. The use of lignocellulose biomass will significantly increase the competitiveness of lactic acid-based polymers compared to conventional petroleum-based plastics. *Lactobacillus* sp. are used extensively in industry for starch-based lactic acid production, the majority of which lack the ability to ferment pentose sugars such as xylose and arabinose (Tsuji, 2002). Although *Lactobacillus pentosus*, *L. brevis*, and *Lactococcus lactis* ferment pentoses to lactic acid, pentoses are metabolized using the phosphoketolase pathway which is inefficient for lactic acid production (Garde et al., 2002; Tanaka et al., 2002). In the phosphoketolase pathway, xylulose 5-phosphate is cleaved to glyceraldehyde 3-phosphate and acetylphosphate. With this pathway, the maximum theoretical yield of lactic acid is limited to one per pentose (0.6 g lactic acid per gram xylose) due to the loss of two carbons to acetic acid. Sugarcane bagasse (Patel et al., 2004), SE wood (Woiciechowski et al., 1999), soybean stalk (Xu et al., 2007), corncob molasses (Wang et al., 2010), trimming vine shoots (Bustos et al., 2007), and wheat straw (Garde et al., 2002) hydrolyzate are used for lactic acid production. John et al. (2006) have reported the production of lactic acid from agro-residues using *Lactobacillus delbrueckii* in solid state and simultaneous saccharification and fermentation.

The US Department of Energy (DOE) listed succinic acid as one of the top 12 chemical building blocks from biomass (US Department of Energy, 2004). Research has been conducted with both fungal and bacterial fermentation. The primary source material is the six-carbon sugar glucose, but some strains will utilize xylose too. Succinic acid has in the past been produced by a specially engineered organism, *A. succiniciproducens*. Research has been conducted to encourage overexpression of succinate in *E. coli*. This strain of *E. coli* was developed by DOE laboratories and subsequently licensed to Bioamber. The DOE technology consumes CO_2 and is reportedly available for further licensing. Succinic acid is a platform chemical, from which a variety of derivatives may be created. Potential uses include green solvents, surfactants/detergents, runway deicer, and a range of other potential food and pharmaceutical uses (Paster et al., 2003). There is little published information about the price of succinic acid. One estimate is that there is a potential 250,000 t/year market at a price of roughly $1.25 per pound.

4.6 BUTANOL

Butanol, an advanced biofuel, is a 4-carbon alcohol (butyl alcohol). It offers a number of advantages and can help accelerate biofuel adoption in countries around the world. It can be produced through processing of domestically-grown crops, such as corn and sugar beets, and other biomass, such as fast-growing grasses and agricultural waste products. Several researchers have studied production of butanol from hydrolyzates of pine, aspen, and corn stover. Biobutanol's primary use is as an industrial solvent in products such as lacquers and enamels and is also compatible with ethanol blending and can improve the blending of ethanol with gasoline. Using fermentation to replace chemical processes in the production of butanol depends largely on the availability of inexpensive and abundant raw materials and efficient conversion of these materials to solvents. Solventogenic acetone butanol ethanol (ABE)-producing *Clostridia* have an added advantage over many other cultures as they can utilize both hexose and pentose sugars (Singh, 1995), which are released from wood and agricultural residues upon hydrolysis, to produce ABE. Parekh et al. (1988) produced ABE from hydrolyzates of pine, aspen, and corn stover using *Clostridium acetobutylicum* P262. Similarly, Marchal et al. (1984) used wheat straw hydrolyzate and *Clostridium acetobutylicum*, while Soni et al. (1982) used bagasse and rice straw hydrolyzates and *Clostridium saccharoperbutylacetonicum* to convert these agricultural wastes into ABE. Nasib Qureshi et al. (2008) have studied the production of butanol from corn fiber hydrolyzate using *Clostridium beijerinckii* BA101. Sun and Liu (2010) have reported the production of butanol from sugar maple hemicellulose hydrolyzate using *Clostridium acetobutylicum* ATCC824.

4.7 BIOHYDROGEN

Hydrogen is considered to be an ideal energy alternative for the future. Compared with the conventional hydrogen generation process (thermochemical and electrochemical), biohydrogen production processes are more environmentally-friendly and less energy intensive. Hydrogen could be produced from renewable materials, such as wastewater, organic wastes, corn straws, and wastewater sludge. Various hemicellulosic hydrolyzates from wheat straw (Kaparaju et al., 2009; Kongjan and Angelidaki, 2010), corn stover (Cao et al., 2009), sugarcane bagasse (Pattra et al., 2008), sweet sorghum (Ivanova et al., 2009), and corn straw (Xu et al., 2010) have been evaluated for biohydrogen production. Biohydrogen has the potential to considerably reduce costs and environmental impact as it can be produced with sunlight and minimal nutrients or organic waste effluents. Hydrogen-producing microorganisms can be rapidly grown in bioreactors with relatively small energy and environmental footprints, making biohydrogen a renewable and low impact technology. Biohydrogen production may provide a renewable, more sustainable alternative but has yet to reach a scale large enough for consideration in replacing a significant portion of the hydrogen supply (Brentner et al., 2010).

4.8 CHITOSAN

Chitosan is an important kind of biomacromolecule. It has protective, and many intermediate, metabolites and energy charges in cells. Chitosan itself has found numerous applications in food, cosmetic, and pharmaceutical industries because of their unique properties such as nontoxicity, biodegradability, biocompatibility, and film-forming and chelating properties along with their antimicrobial activity (Rabea et al., 2003; Silva et al., 2006; Synowiecki and Al-Khateeb, 2003). Tai et al. (2010) have studied the potential of hemicellulose hydrolyzate of corn straw for the production of chitosan by *Rhizopus oryzae*.

4.9 XYLO-OLIGOSACCHARIDES

Xylo-pligosaccharides are xylose-based oligomers that may have variable proportions of substitute groups like acetyl, uronic, and phenolic acids, depending on the lignocellulosic from which they are extracted and the process of production. Substituted xylo-oligosaccharides have some specific characteristics that are driving research efforts to develop applications in fields related to the food and pharmaceutical industries. They may be used as soluble dietary fiber because of its low calorific value and acceptable organoleptic properties. Furthermore, they are noncarcinogenic and act as prebiotics promoting the growth of beneficial *Bifidobacteria* in the colon (Crittenden and Playne, 1996; Yuan et al., 2005) and are a possible ingredient in functional foods (Cummings et al., 2004; Gibson, 2004). Some studies point to the beneficial effect xylo-oligosaccharides may have on reducing the risk of colon cancer (Hsu et al., 2004; Nabarlatz et al., 2007; Wollowski et al., 2001). Recently, xylo-oligosaccharides extracted by autohydrolysis of bamboo have been found to possess a cytotoxic effect on human leukemia cells (Ando et al., 2004). Also xylo-oligosaccharides can be used as a source of xylose for the production of xylitol, a well-known low-calorie sweetener (Rivas et al., 2002).

4.10 FERULIC ACID

Ferulic acid is the major phenolic acid of cereal grains such as wheat, triticale, and rye (Sutherland et al., 1983; Weidner et al., 1999, 2000, 2002). It can exist as an extractable form, as free, esterified, and glycosylated phenolic constituents (Weidner et al., 1999) as well as an insoluble-bound occurring in the outer layers of wheat grains (Kim et al., 2006). Alkaline hydrolysis is reported to release ferulic acid from the insoluble form (Kim et al., 2006; Steinhart and Renger, 2000). Also, enzyme preparations are used for the same purpose (Moore et al., 2006). There are numerous reports on antiradical, oxidase inhibitory, antiinflammatory, antimicrobial, anticancer activities of ferulic acid and its derivatives in the literature data (Hirata et al., 2005; Ou and Kwok, 2004; Wang et al., 2007). Feruloyl esterases (FAEs) act synergistically with xylanases to hydrolyze ester-linked ferulic (FA) and diferulic (diFA) acid from cell wall material and therefore play a major role in the degradation of plant biomass (Topakas et al., 2007). FAEs are used as a tool for the release of ferulic acid from agroindustrial byproducts such as wheat bran, maize bran, maize fiber, brewers, spent grain, sugar beet pulp, coastal Bermuda grass, oat hulls, jojoba meal, wheat straw, coffee pulp, and apple marc (Bartolome et al., 1995, 1997a,b, 2003; Benoit et al., 2006; Borneman et al., 1990; Brezillon et al., 1996; Faulds and Williamson, 1993, 1995; Faulds et al., 1995, 2002, 2003, 2004, 2006; Ferreira et al., 1993, 1999; Kroon and Williamson, 1996; Kroon et al.,

1996, 1999; Laszlo et al., 2006; MacKenzie and Bilous, 1988; Moore et al., 1996; Ralet et al., 1994; Sancho et al., 1999; Shin et al., 2006; Thibault et al., 1998; Topakas and Christakopoulos, 2004; Topakas et al., 2003a,b, 2004, 2005a,b; Yu et al., 2002a,b, 2003).

4.11 VANILLIN

Vanillin can be obtained by fermentation using suitable microorganisms in the stationary growth phase (Converti et al., 2003; De Faveri et al., 2007; Torre et al., 2004a,b). The microbial transformation of ferulic acid is recognized as one of the most attracting alternatives to be able to produce natural vanillin. Bacteria belonging to different genera are able to metabolize ferulic acid as the sole carbon source, producing vanillin, vanillic acid, and protocatechuic acid as catabolic intermediates (Burri et al., 1989; Walton et al., 2003). Vanillin is industrially used as fragrance in food preparations, an intermediate in the productions of herbicides, antifoaming agents or drugs (Davidson and Naidu, 2000), an ingredient of household products such as air fresheners and floor polishes, and, because of its antimicrobial and antioxidant properties (Gould, 1996; Priefert et al., 2001), also as a food preservative (Serra et al., 2005). The much higher price of the natural vanillin compared to the synthetic vanillin has been leading to growing interest of the flavor industry in producing it from natural sources by bioconversion (Davidson and Naidu, 2000; Gurujeyalakshmi and Mahadevan, 1987; Yoon et al., 2005). *Pseudomonas fluorescens* was shown to produce vanillic acid from ferulic acid (Andreoni et al., 1995; Barghini et al., 1998), with formation of vanillin as an intermediate (Narbad and Gasson, 1998). Promising vanillin concentrations were obtained from ferulic acid by *Amycolatopsis* sp. (Gurujeyalakshmi and Mahadevan, 1987; Rabenhost and Hopp, 1997), and *Streptomyces setonii* (Gunnarsson and Palmqvist, 2006; Müller et al., 1998). However, the process development is difficult because of the well-known slow growth of actinomycetes and high viscosity of broths fermented by them; therefore, the construction of new recombinants strains of quickly growing bacteria able to overproduce vanillin is attractive. The biotechnological process to produce vanillin from various agro byproducts had been investigated using different micro-organisms as biocatalysts. *Aspergillus niger* I-1472 and *Pycnoporus cinnabarinus* MUCL39533 were used in a two-step bioconversion using sugar beet pulp (Bonnina et al., 2001; Lesage-Meessen et al., 1999), maize bran (Lesage-Meessen et al., 2002), rice bran oil (Zheng et al., 2007), and wheat bran (Thibault et al., 1998). Wheat bran and corncob have been reported as a good substrate for biovanillin production by *Escherichia coli* JM 109/pBB1 (Di Gioia et al., 2007; Torres et al., 2009).

4.12 FERMENTATION SUBSTRATE FOR PRODUCTION OF ENZYMES

Xylan-containing substrates, and in some instances xylose, can serve as inducers for production of xylan-degrading enzymes including xylanase and xylosidase. Another enzyme of importance that can be produced on these substrates is xylose isomerase (EC 5.3.1.5). This enzyme is used industrially to convert glucose to fructose in the manufacture of high-fructose corn syrup (HFCS) (Bhosale et al., 1996). HFCS is produced by milling corn to produce corn starch that is first treated with alpha-amylase to produce shorter chain oligosaccharides and then with glucoamylase to produce glucose. Finally, xylose isomerase (also known as glucose isomerase) converts glucose to a mixture of about 42% fructose and 50–52% glucose (HFCS-42) with some other sugars mixed in. This 42–43% fructose–glucose mixture is then subjected to a liquid chromatography step, where the fructose is enriched to about 90% and then back-blended with 42% fructose to achieve a 55% fructose final product (HFCS-55). While the relatively inexpensive alpha-amylase and glucoamylase enzymes are added directly to the slurry and used only once, the more costly xylose isomerase is packed into columns and used repeatedly until it loses its activity. Thus, production of HFCS using xylose isomerase is the major application of immobilized enzyme technology (Parker et al., 2010). The most widely used varieties of HFCS are HFCS-55 (mostly used in soft drinks) and HFCS-42 (used in many foods and baked goods). In the USA, HFCS is among the sweeteners that have primarily replaced sucrose. Factors for this include governmental production quotas of domestic sugar, subsidies of US corn, and an import tariff on foreign sugar, all of which combine to raise the price of sucrose to levels above those of the rest of the world, making HFCS less costly for many sweetener applications. Pure fructose is the sweetest of all naturally-occurring carbohydrates and 1.73 times as sweet as sucrose (Hyvonen and Koivistoinen, 1982). Fructose has the lowest GI (19) of all the natural sugars and may be used in moderation by diabetics. In comparison, ordinary table sugar (sucrose) has a GI of 65 and honey has a GI of 55. Per relative sweetness, HFCS-55 is comparable to sucrose. Currently,

HFCS dominate industrial sugar market in the USA. The average American consumed approximately 17.1 kg of HFCS in 2008 versus 21.2 kg of sucrose. In Japan, HFCS consumption accounts for one quarter of total sweetener consumption. The world market for HFCS was 5 million tons in 2004.

4.13 PRODUCTION OF OTHER VALUE-ADDED PRODUCTS

A DOE study (Werpy and Petersen, 2004) has identified the top 12 building blocks that may be produced from sugars. Itaconic acid is also known as methyl succinic acid. It is an unsaturated dicarboxylic organic acid (Willke and Vorlop, 2001). It is a five-carbon sugar with the formula $C_5O_4H_4$ and is one of the 12 building block chemicals identified by DOE. Itaconic acid can be produced by fermentation from C5 and C6 monomers. Biosynthesis of itaconic acid by fungi has been reported as far back as 1932 (Willke and Vorlop, 2001). The chief fungus used is *Aspergillus terreus*. With a glucose substrate (six-carbon sugars), yields are in the range of 40–60%. Yields from five-carbon sugars are in the range of 15–30%. The DOE identified itaconic acid as among the top 12 candidates for chemical production from biomass. Itaconic acid is primarily produced using fermentation and used as a specialty monomer (US Department of Energy, 2004). Itaconic acid has potential for use in the manufacture of acrylic fibers, detergents, adhesives, thickeners, and binders, and a range of other products. In particular, itaconic acid can be incorporated into polymers. It has a potential to be used as a substitute for acrylic or methyl methacrylate, both of which are derived from petrochemicals.

Costs associated with current fermentation processes are a major barrier to the production of itaconic acid. Current processes need to be improved in terms of increased fermentation rate and yield. Increased yield will improve the economics of separation and concentration (US Department of Energy, 2004). Notable suppliers in the USA include Cargill that produces itaconic acid using a corn-based feedstock.

Subsequently, itaconic acid can be converted into polymers through two major routes:

1) First route involves the radical homopolymerization of itaconic acid to polyitaconic acid (Yang and Lu, 2000). Polyitaconic acid is a highly water soluble and highly hydroscopic material and

may be used in paper coating to allow optimal dispersion of the pigment for paper leveling.

2) Second route involves the formation by step polymerization of an unsaturated polyester from itaconic acid and a sugar-derived polyol such as propane diol, butane diol, or methyl butane diol (Werpy and Petersen, 2004). Such polymers are essentially hydrophobic and can react with vinylic monomers such as styrene and methylmethacrylate to produce tough thermosets for usage in structural material such as wood composites and sheet molding compounds.

Conversion of hemicelluloses into polymers of itaconic acid presents a great economic opportunity for an Integrated forest biorefinery (IFBR).

Another example is the production of carbon fibers using lignin precipitated from alkaline hardwood black liquor. Carbon fibers can be made from hardwood kraft lignin when mixed with commercial polymers such as polyesters, polyolefins, and polyethylene oxide (PEO) (Kadla et al., 2002). A main requirement for processing the lignin is that it contains a minimum of volatile compounds, sugars, and ash. Because the actual spinning of the fibers occurs at a temperature of about 220°C, a minimal amount of gaseous components should release at this temperature to avoid bubbles in the fibers and thus lower physical properties and avoid spinning problems. Thus, filtration to remove particulates, carbohydrate stripping, and washing of (almost) sulfur-free lignin will be needed to obtain a suitable feedstock for carbon fiber production (Griffith et al., 2003). The automobile industry would consider large-scale replacement of structural steel in vehicles by composite material containing carbon fibers if the price is decreased to $7,000/MT in order to take advantage of the potential two-third weight reduction. It has been estimated that 10% of the US kraft lignin production is sufficient to produce enough carbon fiber to replace half of the steel in all US passenger vehicles (Griffith et al., 2003). As part of their IFBR efforts, a pilot trial by STFI in STFI-Packforsk Sweden has produced 8 MT of a low-ash precipitated kraft lignin at an estimated cost of about $100/MT (Axegård, 2005). Therefore, conversion of precipitated hardwood kraft lignin to carbon fibers at a yield of 45% means that the lignin feedstock would represent only about 3% of the carbon fiber price. This price margin as well as the large potential market as a structural material in automobiles makes this an attractive product for an IFBR. In addition, the resin needed for imbedding the carbon fibers in the structural panels could be produced from hemicellulose-derived polymers.

A study commissioned by Industry Canada (2007) predicted North American markets for a number of high-growth chemical intermediates, several of them not listed in the DOE report, in 2020. Polylactic acid, whose production is expected to grow by 19% per year to 640,000 MT by 2020, is presently produced by Cargill as a biodegradable bioplastic. Prices are still relatively high, limiting demand; as production increases, prices are likely to drop and demand increase. Citric acid, predicted to grow by 3% per year to 450,000 MT by 2020, is used in food and beverages, and capacity appears well matched to the demand at the present time. Growth for propylene glycol is expected to be 4% per year, reaching 1.3 million tons; competition from lower-priced chemicals may make this a difficult market to enter. Sorbitol is expected to grow by 3% per year to 400,000 t; new capacity would have to compete with existing capacity, which appears to be adequate for the market. Formaldehyde growth, at 3%, is expected to lead to a market of 7.5 million tons. Existing capacity is adequate, but demand is growing in Asia. Finally, 1,4-butanediol, with a 4% growth rate, is expected to grow to 860,000 t. Clearly, more detailed market analysis is needed to fully evaluate these and other opportunities.

REFERENCES

Alriksson, B., Horváth, I.S., Sjöede, A., Nilvebrant, N.O., Jönsson, L.J., 2005. Ammonium hydroxide detoxification of spruce acid hydrolysates. Appl. Biochem. Biotechnol. 121–124, 911–922.

Amidon, T.E., Francis, R., Scott, G.M., Bartholomew, J., Ramarao, B.V., Wood, C.D., 2007. Pulp and pulping processes from an integrated forest biorefinery. Appl. No. PCT/US2005/013216.

Ando, H., Ohba, H., Sakaki, T., Takamine, K., Kamino, Y., Moriwaki, S., 2004. Hot compressed-water decomposed products from bamboo manifest a selective cytotoxicity against acute lymphoblastic leukemia cells. Toxicol. Vitro 18, 765–771.

Andreoni, V., Bernasconi, S., Bestetti, G., 1995. Biotransformation of ferulic acid and related compounds by mutant strains of *Pseudomonas fluorescens*. Appl. Microbiol. Biotechnol. 42, 830–835.

Axegård, P., 2005. The future pulp mill—A biorefinery. In: First International Biorefinery Workshop. Washington DC July 20–21.

Bajpai, P., 2012. Biotechnology in Pulp and Paper Processing. Springer-Verlag, New York, NY.

Barghini, P., Montebove, F., Ruzzi, M., Schiesser, A., 1998. Optimal conditions for bioconversion of ferulic acid into vanillic acid by *Pseudomonas fluorescens* BF13 cells. Appl. Microbiol. Biotechnol. 49, 309–314.

Bartolome, B., Faulds, C.B., Tuohy, M., Hazlewood, G.P., Gilbert, H.J., Williamson, G., 1995. Influence of different xylanases on the activity of ferulic acid esterase on wheat bran. Biotechnol. Appl. Biochem. 22, 65–73.

Bartolome, B., Faulds, C.B., Williamson, G., 1997a. Enzymic release of ferulic acid from barley spent grain. J. Cereal. Sci. 25, 285–288.

Bartolome, B., Faulds, C.B., Kroon, P.A., Waldron, K., Gilbert, H.J., Hazlewood, G., et al., 1997b. An *Aspergillus niger* esterase (ferulic acid esterase III) and a recombinant *Pseudomonas fluorescens* subsp. cellulosa esterase (XylD) released a 5–50 dihydrodimer (diferulic acid) from barley and wheat cell walls. Appl. Environ. Microbiol. 63, 208–212.

Bartolome, B., Gomes-Cordoves, C., Sancho, A.I., Diez, N., Ferreira, P., Soliveri, J., et al., 2003. Growth and release of hydroxycinnamic acids from Brewer's spent grain by *Streptomyces avermitilis* CECT 3339. Enzyme Microb. Technol. 32, 140–144.

Benoit, I., Navarro, D., Marnet, N., Rakotomanomana, N., Lesage-Meessen, L., Sigoillot, J.C., et al., 2006. Feruloyl esterases as a tool for the release of phenolic compounds from agroindustrial by-products. Carbohydr. Res. 341, 1820–1827.

Bhosale, S.H., Mala, B.R., Deshpande, V.V., 1996. Molecular and industrial aspects of glucose isomerase. Microbiol. Rev. 60, 280–300.

Bonnina, E., Brunel, M., Gouy, Y., Lesage-Meessen, L., Asther, M., Thibault, J.F., 2001. *Aspergillus niger* I-1472 and *Pycnoporus cinnabarinus* MUCL39533, selected for the biotransformation of ferulic acid to vanillin, are also able to produce cell wall polysaccharide-degrading enzymes and feruloyl esterases. Enzyme Microb. Technol. 28, 70–80.

Borneman, W.S., Hartley, R.D., Morrison, W.H., Akin, D.E., Ljungdahl, L.G., 1990. Feruloyl and *p*-coumaroyl esterase from anaerobic fungi in relation to plant cell wall degradation. Appl. Microbiol. Biotechnol. 33, 345–351.

Brandberg, T., Franzén, C.J., Gustafsson, L., 2004. The fermentation performance of nine strains of *Saccharomyces cerevisiae* in batch and fed-batch cultures in dilute acid wood hydrolysate. J. Biosci. Bioeng. 98 (2), 122–125.

Brentner, L., Peccia, J., Zimmerman, J., 2010. Challenges in developing biohydrogen as a sustainable energy source: implications for a research agenda. Environ. Sci. Technol. 44, 2243–2254.

Brezillon, C., Kroon, P.A., Faulds, C.B., Brett, G.M., Williamson, G., 1996. Novel ferulic acid esterases are induced by growth of *Aspergillus niger* on sugar-beet pulp. Appl. Microbiol. Biotechnol. 45, 371–376.

Brown, S.F., 2006. Bioplastic fantastic. Fortune 148, 92–94.

Burri, J., Graf, M., Lambelet, P., Löiger, J., 1989. Vanillin: more than a flavouring agent a potent antioxidant. J. Sci. Food Agric. 48, 49–56.

Bustos, G., Torre, N., Moldes, A.B., Cruz, J.M., Domínguez, J.M., 2007. Revalorization of hemicellulosic trimming vine shoots hydrolyzates through continuous production of lactic acid and biosurfactants by *L. pentosus*. J. Food Eng. 78, 405–412.

Canilha, L., Silva, J.B.A., Solenzal, A.I.N., 2004. Eucalyptus hydrolyzate detoxification with activated charcoal adsorption or ion-exchange resins for xylitol production. Process Biochem. 39, 1909–1912.

Cantarella, M., Cantarella, L., Gallifuoco, A., Spera, A., Alfani, F., 2004. Comparison of different detoxification methods for steam-exploded poplar wood as a substrate for the bioproduction of ethanol in SHF and SSF. Process Biochem. 39, 1533–1542.

Cao, G., Ren, N., Wang, A., Lee, D.L., Guo, W., Liu, B., et al., 2009. Acid hydrolysis of corn stover for biohydrogen production using *Thermoanaerobacterium thermosaccharolyticum* W16. Int. J. Hydrogen Energ. 34, 7182–7188.

Carr, F.J., Chill, D., Maida, N., 2002. The lactic acid bacteria: a literature survey. Crit. Rev. Microbiol. 28, 281–370.

Carvalho, W., Silva, S.S., Converti, A., Vitolo, M., 2002. Metabolic behavior of immobilized *Candida guilliermondii* cells during batch xylitol production from sugarcane bagasse acid hydrolyzate. Biotechnol. Bioeng. 79, 165–169.

Chandel, A.K., Kapoor, R.K., Singh, A., Kuhad, R.C., 2007. Detoxification of sugarcane bagasse hydrolyzate improves ethanol production by *Candida shehatae* NCIM 3501. Bioresour. Technol. 98 (10), 1947–1950.

Chen, L.F., Gong, C.S., 1985. Fermentation of sugarcane bagasse hemicellulose hydrolyzate to xylitol by a hydrolyzate acclimatized yeast. J. Food Sci. 50, 226–228.

Chiang, E., Knight, S.G., 1960. Xylose metabolism by cell-free extract of *Penicillium chrysosporium*. Nature 188, 79–81.

Chopin, A., 1993. Organization and regulation of genes for amino acid biosynthesis in lactic acid bacteria. FEMS Microbiol. Rev. 12, 21–38.

Christopher, L., 2012. Adding value prior to pulping: bioproducts from hemicellulose. In: Clement A. Okia (Ed.), Global Perspectives on Sustainable Forest Management. ISBN 978-953-51-0569-5, doi: 10.5772/36849.

Chung, Y.S., Kim, M.D., Lee, W.J., Ryu, Y.W., Kim, J.H., Seo, J.H., 2002. Stable expression of xylose reductase gene enhances xylitol production in recombinant *Saccharomyces cerevisiae*. Enzyme Microb. Technol. 30, 809–816.

Converti, A., Dominguez, J.M., Perego, P., Silva, S.S., Zilli, M., 2000. Wood hydrolysis and hydrolyzate detoxification for subsequent xylitol production. Chem. Eng. Technol. 23, 1013–1020.

Converti, A., De Faveri, D., Perego, P., Barghini, P., Ruzzi, M., Sene, L., 2003. Vanillin production by recombinant strains of *Escherichia coli*. Braz. J. Microbiol. 34, 108–110.

Crittenden, R., Playne, M., 1996. Production, properties, and applications of food-grade oligosaccharides. Trends Food Sci. Tech. 7, 353–361.

Cruz, J.M., Dominguez, J.M., Dominguez, H., Parajo, J.C., 1999. Solvent extraction of hemicellulosic wood hydrolyzates: a procedure useful for obtaining both detoxified fermentation media and polyphenols with antioxidant activity. Food Chem. 67, 147–153.

Cummings, J., Edmond, L., Magee, E., 2004. Dietary carbohydrates on health: do we still need the fibre concept? Clin. Nutr. Suppl. 1, 5–17.

Datta, R., Tsai, S., Bonsignore, P., Moon, S., Frank, J.R., 1995. Technological and economic potential of poly(lactic acid) and lactic acid derivatives. FEMS Microbiol. Rev. 16, 221–231.

Davidson, P.M., Naidu, A.S., 2000. Phytophenols. In: Naidu, A.S. (Ed.), Natural Food Antimicrobial Systems. CRC Press, Boca Raton, FL.

De Faveri, D., Torre, P., Aliakbarian, B., Domínguez, J.M., Perego, P., Converti, A., 2007. Response surface modeling of vanillin production by *Escherichia coli* JM109/pBB1. Biochem. Eng. J. 36, 268–275.

Dibner, J.J., Butin, P., 2002. Use of organic acids as a model to study the impact of gut microflora on nutrition and metabolism. J. Appl. Poultry Res. 11, 453–463.

Di Gioia, D., Sciubba, L., Setti, L., Luziatelli, F., Ruzzi, M., Zanichelli, D., 2007. Production of biovanillin from wheat bran. Enzyme Microb. Technol. 41, 498–505.

Dominguez, J.M., Gong, C.S., Tsao, G.T., 1996. Pretreatment of sugar cane bagasse hemicellulose hydrolyzate for xylitol production by yeast. Appl. Biochem. Biotechnol. 57–58, 49–56.

Faveri, D.D., Perego, P., Converti, A., Borghi, M., 2002a. Xylitol recovery by crystallization from synthetic solutions and fermented hemicellulose hydrolyzates. Chem. Eng. J. 90, 291–298.

Faulds, C.B., Williamson, G., 1993. Ferulic acid esterase from *Aspergillus niger*: purification and partial characterization of two forms from a commercial source of pectinase. Biotechnol. Appl. Biochem. 17, 349–359.

Faulds, C.B., Williamson, G., 1995. Release of ferulic acid from wheat bran by a ferulic acid esterase (FAE III) from *Aspergillus niger*. Appl. Microbiol. Biotechnol. 43, 1082–1087.

Faulds, C.B., Kroon, P.A., Saulnier, L., Thibault, J.F., Williamson, G., 1995. Release of ferulic acid from maize bran and derived oligosaccharides by *Aspergillus niger* esterases. Carbohydr. Polym. 27, 187–190.

Faulds, C.B., Sancho, A.I., Bartolome, B., 2002. Mono- and dimeric ferulic acid release from brewer's spent grain by fungal feruloyl esterases. Appl. Microbiol. Biotechnol. 60, 489–494.

Faulds, C.B., Zanichelli, D., Crepin, V.F., Connerton, I.F., Juge, N., Bhat, M.K., et al., 2003. Specificity of feruloyl esterases for water-extractable and water unextractable feruloylated polysaccharides: influence of xylanase. J. Cereal. Sci. 38, 281–288.

Faulds, C.B., Mandalari, G., LoCurto, R., Bisignano, G., Waldron, K.W., 2004. Arabinoxylan and mono- and dimeric ferulic acid release from brewers grain and wheat bran by feruloyl esterases and glycosyl hydrolases from *Humicola insolens*. Appl. Microbiol. Biotechnol. 64, 644–650.

Faulds, C.B., Mandalari, G., LoCurto, R., Bisignano, G., Christakopoulos, P., Waldron, K.W., 2006. Synergy between xylanases from glycoside hydrolase family 10 and family 11 and a feruloyl esterase in the release of phenolic acids from cereal arabinoxylan. Appl. Microbiol. Biotechnol. 71, 622–629.

Ferreira, L.M.A., Wood, T.M., Williamson, G., Faulds, C.B., Hazlewood, G., Gilbert, H.J., 1993. A modular esterase from *Pseudomonas fluorescens* subsp. cellulose contains a non-catalytic binding domain. Biochem. J. 294, 349–355.

Ferreira, P., Diez, N., Gutierrez, C., Soliveri, J., Copa-Patino, J.L., 1999. *Streptomyces avermitilis* CECT 3339 produces a ferulic acid esterase able to release ferulic acid from sugar beet pulp soluble feruloylated oligosaccharides. J. Sci. Food Agric. 79, 440–442.

Garde, A., Jonsson, G., Schmidt, A.S., Ahring, B.K., 2002. Lactic acid production from wheat straw hemicellulose hydrolysate by *Lactobacillus pentosus* and *Lactobacillus brevis*. Bioresour. Technol. 81, 217–223.

Garg, S.K., Jain, A., 1995. Fermentative production of 2,3-butanediol. Bioresour. Technol. 51, 103–109.

Gibson, G., 2004. Fibre and effects on probiotics (the prebiotic concept). Clin. Nutr. Suppl. 1, 25–31.

Gong, C.S., Cao, N., Tsao, G.T., 1997. Biological production of 2,3-butanediol from renewable biomass. In: Saha, B.C., Woodward, J. (Eds.), Fuels and Chemicals from Biomass. American Chemical Society, Washington, DC, pp. 280–293.

Gonzalez, J., Cruz, J.M., Dominguez, H., Parajo, J.C., 2004. Production of antioxidants from *Eucalyptus globulus* wood by solvent extraction of hemicellulose hydrolyzates. Food Chem. 84, 243–251.

Gould, G.W., 1996. Industry perspectives on the use of natural antimicrobials and inhibitors for food applications. J. Food Prot. 13, 82–86.

Govinden, R., Pillav, B., van Zyl, W.H., Pillay, D., 2001. Xylitol production by recombinant *Saccharomyces cerevisiae* expressing the *Pichia stipitis* and *Candida shehatae* XYLI genes. Appl. Microbiol. Biotechnol. 55, 76–80.

Granstorm, T., Ojama, H., Leisola, M., 2001. Chemostat study of xylitol production by *Candida guilliermondii*. Appl. Microbiol. Biotechnol. 55, 36–42.

Griffith, W.L., Compere, A.L., Leitten, C.F., Shaffer, J.T., 2003. Low-cost, lignin-based carbon fiber for transportation applications. Int. Samp. Tech. Conf. 35, 142–149.

Gunnarsson, N., Palmqvist, E.A., 2006. Influence of pH and carbon source on the production of vanillin from ferulic acid by *Streptomyces setonii* ATCC 39116. Develop. Food Sci. 43, 73–76.

Gurgel, P.V., Manchilha, I.M., Pecanha, R.P., Siqueira, J.F.M., 1995. Xylitol recovery from fermented sugarcane bagasse hydrolyzate. Bioresour. Technol. 52, 219–223.

Gurujeyalakshmi, G., Mahadevan, A., 1987. Dissimilation of ferulic acid by *Bacillus subtilis*. Curr. Microbiol. 16, 69–73.

Gutiérrez, T., Ingram, L.O., Preston, J.F., 2006. Purification and characterization of a furfural reductase (FFR) from *Escherichia coli* strain LYO1—an enzyme important in the detoxification of furfural during ethanol production. J. Biotechnol. 121, 154–164.

Hallborn, J., Walfridsson, M., Airaksine, U., Ojamo, H., Hahn-Hägerdal, B., 1991. Xylitol production by recombinant *Saccharomyces cerevisiae*. Biotechnology 9, 1090–1095.

Hallborn, J., Gorwa, M.F., Meinander, N., Penttila, M., Keranen, S., Hahn-Hägerdal, B., 1994. The influence of cosubstrate and aeration on xylitol formation by recombinant *Saccharomyces cerevisiae* expressing the XYLI gene. Appl. Microbiol. Biotechnol. 42, 326–333.

Hirata, A., Murakami, Y., Atsumi, T., Shoji, M., Ogiwara, T., Shibuya, K., et al., 2005. Ferulic acid dimer inhibits lipopolysaccharide-stimulated cyclooxygenase-2 expression in macrophages. In Vivo 19, 849–853.

Hofer, M., Betz, A., Kotvk, A., 1971. Metabolism of the obligatory aerobic waste *Rhodotorula gracilus* IV. Induction of an enzyme necessary for xylose catabolism. Biochim. Biophys. Acta 252, 1–12.

Hofyendahl, K., Hahn-Hägerdal, B., 2000. Factors affecting the fermentative lactic acid production from renewable resources. Enzyme Microb. Technol. 26, 87–107.

Horváth, I.S., Sjoede, A., Alriksson, B., Jönsson, L.J., Nilvebrant, N.O., 2005. Critical conditions for improved fermentability during overliming of acid hydrolysates from spruce. Appl. Biochem. Biotechnol. 121–124, 1031–1044.

Hsu, C.K., Liao, J.W., Chung, Y.C., Hsieh, C.P., Chan, Y.C., 2004. Xylooligosaccharides and fructooligosaccharides affect the intestinal microbiota and precancerous colonic lesion development in rats. J. Nutr. 134, 1523–1528.

Hyvonen, L., Koivistoinen, P., 1982. Fructose in food systems. In: Birch, G.G., Parker, K.J. (Eds.), Nutritive Sweeteners. Applied Science Publishers, London, pp. 133–144.

Industry Canada, 2007. Towards a technology roadmap for Canadian forest biorefineries—a report.

Ivanova, G., Rákhely, G., Kovács, K.L., 2009. Thermophilic biohydrogen production from energy plants by *Caldicellulosiruptor saccharolyticus* and comparison with related studies. Int. J. Hydrogen Energ. 34, 3659–3670.

Jansen, N.B., Flickinger, M.C., Tsao, G.T., 1984. Production of 2,3-butanediol from D-xylose by *Klebsiella oxytoca* ATCC 8724. Biotechnol. Bioeng. 26, 362–368.

John, R.P., Nampoothiri, M., Pandey, A., 2006. Solid state fermentation for lactic acid production from agro waste using *Lactobacillus delbrueckii*. Process Biochem. 41, 759–763.

Jönsson, L.J., Palmqvist, E., Nilvebrant, N.O., Hahn-Hägerdal, B., 1998. Detoxification of wood hydrolyzates with laccase and peroxidase from the white-rot fungus *Trametes versicolor*. Appl. Microbiol. Biotechnol. 49, 691–697.

Kadla, J.F., Kubo, S., Venditti, R.A., Gilbert, R.D., Compere, A.L., Griffith, W., 2002. Lignin-based carbon fibers for composite fiber applications. Carbon 40, 2913–2920.

Kaparaju, P., Serrano, M., Thomsen, A.B., Kongjan, P., Angelidaki, I., 2009. Bioethanol, biohydrogen, biogas production from wheat straw in a biorefinery concept. Bioresour. Technol. 100, 2562–2568.

Karimkulova, M.P., Khakimov, Y.S., Abidova, M.F., 1989. Activity of modified catalysts in hydrogenation of xylose to xylitol. Chem. Nat. Comp. 25, 370–371.

Kim, K.H., 2005. Two-stage dilute acid-catalyzed hydrolytic conversion of softwood sawdust into sugars fermentable by ethanologenic microorganisms. J. Sci. Food Agr. 85 (14), 2461–2467.

Kim, K.H., Tsao, R., Yang, R., Cui, S.W., 2006. Phenolic acid profiles and antioxidant activities of wheat bran extracts and the effect of hydrolysis conditions. Food Chem. 95, 466–473.

Klinke, H.B., Thomsen, A.B., Ahring, B.K., 2004. Inhibition of ethanol-producing yeast and bacteria by degradation products produced during pre-treatment of biomass. Appl. Microbiol. Biotechnol. 66 (1), 10–26.

Kongjan, P., Angelidaki, I., 2010. Extreme thermophilic biohydrogen production from wheat straw hydrolysate using mixed culture fermentation: effect of reactor configuration. Bioresour. Technol. 101, 7789–7796.

Kosaric, N., Velikonja, J., 1995. Liquid and gaseous fuels from biotechnology: challenges and opportunities. FEMS Microbiol. Rev. 16, 111–142.

Kroon, P.A., Williamson, G., 1996. Release of ferulic acid from sugar-beet pulp by using arabinanase, arabinofuranosidase and an esterase from *Aspergillus niger*. Biotechnol. Appl. Biochem. 23, 263–267.

Kroon, P.A., Faulds, C.B., Williamson, G., 1996. Purification and characterisation of a novel esterase induced by growth of *Aspergillus niger* on sugar-beet pulp. Biotechnol. Appl. Biochem. 23, 255–262.

Kroon, P.A., Garcia-Conesa, M.T., Fillingham, I.J., Hazlewood, G.P., Williamson, G., 1999. Release of ferulic acid dehydrodimers from plant cell walls by feruloyl esterases. J. Sci. Food Agric. 79, 428–434.

Kuyper, M., Hartog, M.M.P., Toirkens, M.J., Almering, M.J.H., Winkler, A.A., van Dijken, J.P., et al., 2005a. Metabolic engineering of a xylose-isomerase-expressing *Saccharomyces cerevisiae* strain for rapid anaerobic xylose fermentation. FEMS Yeast Res. 5 (4–5), 399–409.

Kuyper, M., Toirkens, M.J., Diderich, J.A., Winkler, A.A., van Dijken, J.P., Pronk, J.T., 2005b. Evolutionary engineering of mixed-sugar utilization by a xylose-fermenting *Saccharomyces cerevisiae* strain. FEMS Yeast Res. 5 (10), 925–934.

Larsson, S., Palmqvist, E., Hahn-Hägerdal, B., Tengborg, C., Stenberg, K., Zacchi, G., et al., 1999. The generation of fermentation inhibitors during dilute acid hydrolysis of softwood. Enzyme Microb. Tech. 24 (3–4), 151–159.

Laszlo, J.A., Compton, D.L., Li, X.L., 2006. Feruloyl esterases hydrolysis and recovery of ferulic acid from jojoba meal. Ind. Crop. Prod. 23, 46–53.

Lee, W.J., Ryu, Y.W., Seo, J.H., 2000. Characterization of two substrate fermentation processes for xylitol production using recombinant *Saccharomyces cerevisiae* containing xylose reductase. Process Biochem. 35, 1199–1203.

Lesage-Meessen, L., Stentelaire, C., Lomascolo, A., Couteau, D., Asther, M., Moukha, S., 1999. Fungal transformation of ferulic acid from sugar beet pulp to natural vanillin. J. Sci. Food Agric. 79, 487–490.

Lesage-Meessen, L., Lomascolo, A., Bonnin, E., Thibault, J.F., Buleon, A., Roller, M., 2002. A biotechnological process involving filamentous fungi to produce natural crystalline vanillin from maize bran. Appl. Biochem. Biotechnol. 102–103, 141–153.

MacKenzie, C.R., Bilous, D., 1988. Ferulic acid esterase activity from *Schizophyllum commune*. Appl. Environ. Microbiol. 54, 1170–1173.

Maddox, I.S., 1996. Microbial production of 2,3-butanediol. In: Roehr, M. (Ed.), Biotechnology. Vol. 6. Products of Primary Metabolism. VCH, Weinheim, pp. 269–291.

Magee, R.J., Kosaric, N., 1987. The microbial production of 2,3-butanediol. Adv. Appl. Microbiol. 32, 89–161.

Magnuson, J.K., Lasure, L.L., 2004. Organic acid production by filamentous fungi. In: Tkacs, J.S., Lange, L. (Eds.), Advances in Fungal Biotechnology for Industry, Agriculture, and Medicine. Kluwer Academic/Plenum Publishers, New York, NY, pp. 307–340.

Malaja, A., Hamalainen, L., 1977. Process for making xylitol. US Patent 4.008.285.

Mamman, A.S., Lee, J.M., Kim, Y.C., Hwang, I.T., Park, N.J., Hwang, Y.K., et al., 2008. Furfural: Hemicellulose/xylose derived biochemical. Biofpr 5, 438–454.

Marchal, R., Rebeller, M., Vandecasteele, J.P., 1984. Direct bioconversion of alkali pretreated straw using simultaneous enzymatic hydrolysis and acetone butanol production. Biotechnol. Lett. 6, 523–528.

Marinova, M., Mateos-Espejel, E., Paris, J., 2010. From kraft mill to forest biorefinery: an energy and water perspective II—case study. Cellulose Chem. Technol. 44 (1–3), 21–26.

Martin, C., Galbe, M., Wahlbom, C.F., Hahn-Hägerdal, B., Jönsson, L.J., 2002. Ethanol production from enzymatic hydrolyzates of sugarcane bagasse using recombinant xylose-utilising *Saccharomyces cerevisiae*. Enzyme Microb. Technol. 31, 274–282.

Mattey, M., 1992. The production of organic acids. Crit. Rev. Biotechnol. 12, 87–132.

Mattila, P.T., Svanberg, M.J., Jämsä, T., Knuuttila, M.L., 2002. Improved bone biomechanical properties in xylitol-fed aged rats. Metabolism 51, 92–96.

Menon, V., Prakash, G., Rao, M., 2010. Value added products from hemicelluloses: biotechnological perspective. Global J. Biochem. 1 (1), 36–67.

Meinander, N., Hahn-Hägerdal, B., Linko, M., Linko, P., Ojamo, H., 1994. Fed-batch xylitol production with recombinant XYL-I-expressing *Saccharomyces cerevisiae* using ethanol as a co-substrate. Appl. Microbiol. Biotechnol. 42, 334–339.

Mikkola, J.P., Salmi, T., Sjöholm, R., Mäki-Arvela, P., Vainio, H., 2000. Hydrogenation of xylose to xylitol: three-phase catalysis by promoted Raney nickel, catalyst deactivation and *in situ* sonochemical catalyst rejuvenation. Stud. Surf. Sci. Catal. 20, 27–32.

Milgrom, P., Ly, K.A., Roberts, M.C., Rothen, M., Mueller, G., Yamaguchi, D.K., 2006. Mutans streptococci dose response to xylitol chewing gum. J. Dent. Res. 85, 177–181.

Millati, R., Edebo, L., Taherzadeh, M.J., 2005. Performance of *Rhizopus, Rhizomucor*, and *Mucor* in ethanol production from glucose, xylose, and wood hydrolyzates. Enzyme Microb. Tech. 36 (2–3), 294–300.

Moore, J., Bamforth, C.W., Kroon, P.A., Bartolome, B., Williamson, G., 1996. Ferulic acid esterase catalyses the solubilization of β-glucans, and pentosans from the starchy endosperm cell walls of barley. Biotechnol. Lett. 18, 1423–1426.

Moore, J., Cheng, Z.H., Su, L., Yu, L.L.L., 2006. Effects of solid-state enzymatic treatments on the antioxidant properties of wheat bran. J. Agric. Food Chem. 54, 9032–9045.

Müller, B., Münch, T., Muheim, A.,Wetli, M., 1998. Process for the preparation of vanillin. European Patent 0885968.

Nabarlatz, D., Ebringerová, A., Montané, D., 2007. Autohydrolysis of agricultural byproducts for the production of xylooligosaccharides. Carbohyd. Polym. 69, 20–28.

Narbad, A., Gasson, M.J., 1998. Metabolism of ferulic acid to vanillin using a novel CoAdependent pathway in a newly isolated strain of *Pseudomonas fluorescens*. Microbiology 144, 1397–1405.

Nasib Qureshi, N., Ezeji, T.C., Ebener, J., Dien, B.S., Cotta, M.A., Blaschek, H.P., 2008. Butanol production by *Clostridium beijerinckii*. Part I: use of acid and enzyme hydrolyzed corn fiber. Bioresour. Technol. 99, 5915–5922.

Nguyen, Q.A., Tucker, M.P., Keller, F.A., Eddy, F.P., 2000. Two-stage dilute-acid pretreatment of softwoods. Appl. Biochem. Biotechnol. 84–86, 561–576.

Nolleau, V., Preziosi-Belloy, L., Navarro, J.M., 1995. The reduction of xylose to xylitol by *Candida guilliermondii* and *Candida parapsilosis*: incidence of oxygen and pH. Biotechnol. Lett. 17, 417–422.

Okano, K., Yoshida, S., Yamada, R., Tanaka, T., Ogino, C., Fukuda, H., et al., 2009. Improved production of homo-D-lactic acid via xylose fermentation by introduction of xylose assimilation genes and redirection of the phosphoketolase pathway to the pentose phosphate pathway in L-lactate dehydrogenase gene-deficient *Lactobacillus plantarum*. Appl. Environ. Microbiol. 75, 7858–7861.

Ou, S.Y., Kwok, K.C., 2004. Ferulic acid: pharmaceutical functions, preparation and applications in foods. J. Sci. Food Agr. 84, 1261–1269.

Palmqvist, E., Hahn-Hägerdal, B., 2000. Fermentation of lignocellulosic hydrolysates, I: inhibition and detoxification. Bioresour. Technol. 74 (1), 17–24.

Parajo, J.C., Dominguez, H., Dominguez, J.M., 1996. Charcoal adsorption of wood hydrolyzates for improving their fermentability: influence. Bioresour. Technol. 57, 179–185.

Parajo, J.E., Dominguez, H., Dominguez, J.M., 1997. Improved xylitol production with *Debaryomyces hansenii* Y-7426 from raw or detoxified wood hydrolyzates. Enzyme Microb. Technol. 21, 18–24.

Parekh, S.R., Parekh, R.S., Wayman, M., 1988. Ethanol and butanol production by fermentation of enzymatically saccharified SO_2-prehydrolysed lignocellulosics. Enzyme Microb. Technol. 10, 660–668.

Parker, K., Salas, M., Nwosu, V.C., 2010. High fructose corn syrup: production, uses and public health concerns. Biotechnol. Mol. Biol. Rev. 5, 71–78.

Paster, M., Pellegrino, J.L., Carole T.M., 2003. Industrial bioproducts: today and tomorrow. Energetic, Incorporated, 86pp.

Patel, M., Ou, M., Ingram, L.O., Shanmugam, K.T., 2004. Fermentation of sugar cane bagasse hemicellulose hydrolysate to L(+)-lactic acid by a thermotolerant acidophilic *Bacillus* sp. Biotechnol. Lett. 26, 865–868.

Pattra, S., Sangyoka, S., Boonmee, M., Reungsang, A., 2008. Bio-hydrogen production from the fermentation of sugarcane bagasse hydrolysate by *Clostridium butyricum*. Int. J. Hydrogen Energ. 33, 5256–5265.

Persson, P., Larsson, S., Jönsson, L.J., Nilvebrant, N.O., Sivik, B., Munteanu, F., et al., 2002. Supercritical fluid extraction of a lignocellulosic hydrolysate of spruce for detoxification and to facilitate analysis of inhibitors. Biotechnol. Bioeng. 79 (6), 694–700.

Pfeifer, M.J., Silva, S.S., Felipe, M.G.A., Roberto, I.C., Mancilha, I.M., 1996. Effect of culture conditions on xylitol production by *Candida guilliermondii* FTI 20037. Appl. Biochem. Biotechnol. 57–58, 423–430.

Prakasham, R.S., Sreenivas, R.R., Hobbs, P.J., 2009. Current trends in biotechnological production of xylitol and future prospects. Curr. Trends Biotechnol. Pharm. 3, 8–36.

Priefert, H., Rabenhorst, J., Steinbüchel, A., 2001. Biotechnological production of vanillin. Appl. Microbiol. Biotechnol. 56, 296–314.

Rabea, E.I., Badawy, M.E., Stevens, C.V., Smagghe, G., Steurbaut, W., 2003. Chitosan as anti-microbial agent: applications and mode of action. Biomacromolecules 4, 1457–1465.

Rabenhost, J., Hopp, R., 1997. Process for the preparation of vanillin and suitable microorganisms. European Patent 0761817.

Ragauskas, A.J., Nagy, M., Kim, D.H., Eckert, C.A., Hallett, J.P., Liotta, C.L., 2006. From wood to fuels: integrating biofuels and pulp production. Indus. Biotech. 2 (1), 55–65.

Ralet, M.C., Faulds, C.B., Williamson, G., Thibault, J.F., 1994. Degradation of feruloylated oligosaccharides from sugar-beet pulp and wheat bran by ferulic acid esterase from *Aspergillus niger*. Carbohydr. Res. 263, 257–269.

Rivas, B., Domínguez, J., Domínguez, H., Parajó, J., 2002. Bioconversion of post-hydrolysed autohydrolysis liquors: an alternative for xylitol production from corncobs. Enzyme Microb. Technol. 31, 431–438.

Saha, B.C., 2003. Hemicellulose bioconversion. J. Ind. Microbiol. Biotechnol. 30, 279–291.

Saha, B.C., Bothast, R.J., 1997. Microbial production of xylitol. In: Saha, B.C., Woodward, J. (Eds.), Fuels and Chemicals from Biomass. American Chemical Society, Washington, DC, pp. 307–319.

Saha, B.C., Bothast, R.J., 1999. Production of 2,3-butanediol by a newly isolated *Enterobacter cloacae*. Appl. Microbiol. Biotechnol. 52, 321−326.

Sancho, A.I., Faulds, C.B., Bartolome, B., Williamson, G., 1999. Characterization of feruloyl esterase activity in barley. J. Sci. Food Agric. 79, 447−449.

Senthilkumar, V., Gunasekaran, P., 2005. Bioethanol production from cellulosic substrates: engineered bacteria and process integration challenges. J. Sci. Ind. Res. 64 (11), 845−853.

Serra, S., Fuganti, C., Brenna, E., 2005. Biocatalytic preparation of natural flavours and fragrances. Trends Biotechnol. 23, 193−198.

Shin, H.D., McClendon, S., Le, T., Taylor, F., Chen, R.R., 2006. A complete enzymatic recovery of ferulic acid from corn residues with extracellular enzymes from *Neosartorya spinosa* NRRL 185. Biotechnol. Bioeng. 95, 1108−1115.

Silva, S.S., Felipe, G.A., Mancilha, I.M., 1998. Factors that affect the biosynthesis of xylitol by xylose-fermenting yeasts. A review. Appl. Biochem. Biotechnol. 70−72, 331−339.

Silva, H.S.R.C., dos Santos, K.S.C.R., Ferreira, E.I., 2006. Chitosan: hydrosoluble derivatives, pharmaceutical applications and recent advances. Quim. Nova. 29, 776−785.

Singh, A., 1995. Microbial production of acetone and butanol. In: Singh, A., Mishra, P. (Eds.), Microbial Pentose Utilization—Current Applications in Biotechnology. Elsevier Science, New York, NY, pp. 197−220.

Soni, B.K., Das, K., Ghose, T.K., 1982. Bioconversion of agro-wastes into acetone butanol. Biotechnol. Lett. 4, 19−22.

Sreenath, H.K., Jeffries, T.W., 1999. Production of ethanol from wood hydrolyzate by yeasts. Bioresour. Technol. 72 (3), 253−260.

Steinhart, H., Renger, A., 2000. Influence of technological processes on the chemical structure of cereal dietary fiber. Czech J. Food Sci. 8, 22−24.

Sun, Z., Liu, S., 2010. Production of *n*-butanol from concentrated sugar maple hemicellulosic hydrolysate by *Clostridium acetobutylicum* ATCC824. Biomass Bioenerg., 1−9.

Sutherland, J.B., Crawford, D.L., Pometto, A.L., 1983. Metabolism of cinnamic, *p*-coumaric and ferulic acids by *Streptomyces setonii*. Can. J. Microbiol. 29, 1253−1257.

Synowiecki, J., Al-Khateeb, N.A., 2003. Production, properties, and some new applications of chitin and its derivatives. Crit. Rev. Food Sci. Nutr. 43, 145−171.

Syu, M.J., 2001. Biological production of 2,3-butanediol. Appl. Microbiol. Biotechnol. 55, 10−18.

Taherzadeh, M.J., Eklund, R., Gustafsson, L., Niklasson, C., Lidén, G., 1997. Characterization and fermentation of dilute-acid hydrolyzates from wood. Ind. Eng. Chem. Res. 36 (11), 4659−4665.

Taherzadeh, M.J., Gustafsson, L., Niklasson, C., Lidén, G., 2000a. Physiological effects of 5-hydroxymethylfurfural on *Saccharomyces cerevisiae*. Appl. Microbiol. Biotechnol. 53 (6), 701−708.

Taherzadeh, M.J., Gustafsson, L., Niklasson, C., Lidén, G., 2000b. Inhibition effects of furfural on aerobic batch cultivation of *Saccharomyces cerevisiae* growing on ethanol and/or acetic acid. J. Biosci. Bioeng. 90 (4), 374−380.

Tai, C., Li, S., Xu, Q., Ying, H., Huang, H., Ouyang, P., 2010. Chitosan production from hemicellulose hydrolysate of corn straw: impact of degradation products on *Rhizopus oryzae* growth and chitosan fermentation. Lett. Appl. Microbiol. 51, 278−284.

Tanaka, K., Komiyama, A., Sonomoto, K., Ishizaki, A., Hall, S.J., Stanbury, P.E., 2002. Two different pathways for D-xylose metabolism and the effect of xylose concentration on the yield coefficient of lactate in mixed-acid fermentation by the lactic acid bacterium *Lactococcus lactis* IO-1. Appl. Microbiol. Biotechnol. 60, 160−167.

Thibault, J.F., Asther, M., Ceccaldi, B.C., Couteau, D., Delattre, M., Duarte, J.C., et al., 1998. Fungal bioconversion of agricultural by-products to vanillin. Lebensm. Wiss. Technol. 31, 530–536.

Thorp, B., 2005. Transition of mills to biorefinery model creates new profit streams. Pulp and Paper November, 35–39.

Thorp, B., Raymond, D., 2005. Forest biorefinery could open door to bright future for P&P industry. Paper Age 120 (7), 16–18.

Tolan, J.S., 2003. Conversion of cellulosic biomass to ethanol using enzymatic hydrolysis. In: 226th American Chemical Society National Meeting Abstracts, New York, NY.

Topakas, E., Christakopoulos, P., 2004. Enzymic release of phenolic antioxidants from plant cell wall material. NutrCos 1-2, 54–57.

Topakas, E., Stamatis, H., Biely, P., Kekos, D., Macris, B.J., Christakopoulos, P., 2003a. Purification and characterization of a feruloyl esterase from *Fusarium oxysporum* catalyzing esterification of phenolic acids in ternary water organic solvent mixtures. J. Biotechnol. 102, 33–44.

Topakas, E., Stamatis, H., Mastihubova, M., Biely, P., Kekos, D., Macris, B.J., et al., 2003b. Purification and characterization of a *Fusarium oxysporum* feruloyl esterase (FoFAE-I) catalyzing transesterification of phenolic acid esters. Enzyme Microb. Technol. 33, 729–737.

Topakas, E., Stamatis, H., Biely, P., Christakopoulos, P., 2004. Purification and characterization of a type B feruloyl esterase (StFAE-A) from the thermophilic fungus *Sporotrichum thermophile*. Appl. Microbiol. Biotechnol. 63, 686–690.

Topakas, E., Christakopoulos, P., Faulds, C.B., 2005a. Comparison of mesophilic and thermophilic feruloyl esterases: characterization of their substrate specificity for methyl phenylalkanoates. J. Biotechnol. 115, 355–366.

Topakas, E., Vafiadi, C., Stamatis, H., Christakopoulos, P., 2005b. *Sporotrichum thermophile* type C feruloyl esterase (StFaeC): purification characterization, and its use for phenolic acid (sugar) ester synthesis. Enzyme Microb. Technol. 36, 729–736.

Topakas, E., Vafiadi, C., Christakopoulos, P., 2007. Microbial production, characterization and applications of feruloyl esterases. Process Biochem. 42, 497–509.

Torre, P., De Faveri, D., Perego, P., Converti, A., Barghini, P., Ruzzi, M., 2004a. Selection of co-substrate and aeration conditions for vanillin production by *Escherichia coli* JM109/pBB1. Food Technol. Biotechnol. 42, 193–196.

Torre, P., De Faveri, D., Perego, P., Ruzzi, M., Barghini, P., Gandolfi, R., 2004b. Bioconversion of ferulate into vanillin by *Escherichia coli* strain JM109/pBB1 in an immobilized cell reactor. Ann. Microbiol. 54, 517–527.

Torres, B.R., Aliakbariana, B., Torrea, P., Peregoa, P., Domínguezb, J.M., Zilli, M., et al., 2009. Vanillin bioproduction from alkaline hydrolyzate of corn cob by *Escherichia coli* JM109/pBB1. Enzyme Microb. Technol. 44, 154–158.

Trahan, L., 1995. Xylitol: a review of its action on mutans streptococci and dental plaque—its clinical significance. Int. Dent. J. 45, 77–92.

Tran, A.V., Chambers, R.P., 1987. The dehydration of fermentative 2,3-butanediol into methyl ethyl ketone. Biotechnol. Bioeng. 29, 343–351.

Tsuji, F., 2002. Autocatalytic hydrolysis of amorphous-made polylactides: effects of lactide content, tacticity, and enantiomeric polymer blending. Polymer 43, 1789–1796.

Uhari, M., Kontiokari, T., Niemelä, M., 1998. A novel use of xylitol sugar in preventing acute otitis media. Pediatrics 102, 879–974.

US Department of Energy, 2004. Top value added chemicals from biomass. Volume 1—Results from screening for potential candidates from sugars and synthesis gas, pp. 67.

US Department of Energy, 2006. Industrial Technologies Program–Hemicellulose Extraction and its integration in pulp production. Available at: http://www1.eere.energy.gov/manufacturing/resources/forest/pdfs/hemicellulose_extraction.pdf.

Van Dijken, J.P., Scheffers, W.A., 1986. Redox balances in the metabolism of sugars by yeasts. FEMS Microbiol. Rev. 32, 199–224.

Van Zyi, C., Prior, B.A., du Preez, J.C., 1991. Acetic acid inhibition of D-xylose fermentation by *Pichia stipitis*. Enzyme Microb. Technol. 13, 82–86.

Villarreal, M.L.M., Prata, A.M.R., Felipe, M.G.A., Silva, J.B.A., 2006. Detoxification procedures of Eucalyptus hemicellulose hydrolyzate for xylitol production by *Candida guilliermondii*. Enzyme Microb. Technol. 40 (1), 17–24.

Walton, N.J., Mayer, M.J., Narbad, A., 2003. Vanillin. Phytochemistry 63, 505–515.

Wang, F., Yang, L.X., Huang, K.X., Li, X.K., Hao, X.J., Stockigt, J., et al., 2007. Preparation of ferulic acid derivatives and evaluation of their xanthine oxidase inhibition activity. Nat. Prod. Res. 21, 196–202.

Wang, L., Zhao, B., Liu, B., Yu, B., Ma, C., Su, F., et al., 2010. Efficient production of L-lactic acid from corncob molasses, a waste by-product in xylitol production, by a newly isolated xylose utilizing *Bacillus* sp. strain. Bioresour. Technol. 101, 7908–7915.

Weidner, S., Amarowicz, R., Karamać, M., Dąbrowski, G., 1999. Phenolic acids in caryopses of two cultivars of wheat, rye and triticale that display different resistance to preharvest sprouting. Eur. Food Res. Tech. 210, 109–113.

Weidner, S., Amarowicz, R., Karamać, M., Frączek, E., 2000. Changes in endogenous phenolic acids during development of *Secale cereale* caryopses and after dehydration treatment of unripe rye grains. Plant Physiol. Biochem. 38, 595–602.

Weidner, S., Krupa, U., Amarowicz, R., Karamać, M., Abe, S., 2002. Phenolic compounds in embryos of triticale caryopses at different stages of development and maturation in normal environment and after dehydration treatment. Euphytica 126, 115–122.

Werpy, T., Petersen, G., 2004. Top Value-Added Chemicals from Biomass, Volume I: Results of Screening for Potential Candidates from Sugars and Synthesis Gas. Pacific Northwest National Laboratory, <http://www.eere.energy.gov/biomass/pdfs/35523.pdf>.

Willke, T., Vorlop, K.D., 2001. Biotechnological production of itaconic acid. Appl. Microbiol. Biotechnol. 56, 289–295.

Wilson, J.J., Deschatelets, L., Nishikawa, N.K., 1989. Comparative fermentability of enzymatic and acid hydrolyzates of steam-pretreated aspenwood hemicellulose by *Pichia stipitis* CBS 5776. Appl. Microbiol. Biotechnol. 31, 592–596.

Win, D.T., 2005. Furfural—gold from garbage. AU J. Technol. 58, 185–190.

Winkelhausen, E., Kuzmanova, S., 1998. Microbial conversion of D-xylose to xylitol. J. Ferment. Bioeng. 86, 1–14.

Woiciechowski, A.L., Soccol, C.R., Ramos, L.P., Pandey, A., 1999. Experimental design to enhance the production of L-(+)-lactic acid from steam-exploded wood hydrolysate using *Rhizopus oryzae* in a mixed-acid fermentation. Proc. Biochem. 34, 949–955.

Wollowski, I., Rechkemmer, G., Pool-Zobel, B., 2001. Protective role of probiotics and prebiotics in colon cancer. Am. J. Clin. Nutr. 73, 451–455.

Wright, J.D., Power, A.J., 1987. Comparative technical evaluation of acid hydrolysis processes for conversion of cellulose to alcohol. Energy Biomass Wastes, 949–971.

Wyman, C.E., Goodman, B.J., 1993. Biotechnology for production of fuels, chemicals, and materials from biomass. Appl. Biochem. Biotechnol. 39-40, 41–59.

Xu, Z., Wang, Q., Wang, P., Cheng, G., Ji, Y., Jiang, Z., 2007. Production of lactic acid from soybean stalk hydrolysate with *Lactobacillus sakei* and *Lactobacillus casei*. Proc. Biochem. 42, 89–92.

Xu, J.F., Ren Ng, D.X., Qiu, J., 2010. Biohydrogen production from acetic acid steam-exploded corn straws by simultaneous saccharification and fermentation with *Ethanoligenens harbinense* B49. Int. J. Hydrogen Energy 34, 381−386.

Yang, C.Q., Lu, Y., 2000. Text Res. J. 70 (4), 359−362.

Yoon, S., Li, C., Kim, J., Lee, S., Yoon, J., Choi, M., 2005. Production of vanillin by metabolically engineered *Escherichia coli*. Biotechnol. Lett. 27, 1829−1832.

Yu, E.K.C., Saddler, I.N., 1985. Biomass conversion to butanediol by simultaneous saccharification and fermentation. Trends Biotechnol. 3, 100−104.

Yu, P., Maenz, D.D., McKinnon, J.J., Racz, V.J., Christensen, D.A., 2002a. Release of ferulic acid from oat hulls by *Aspergillus* ferulic acid esterase and *Trichoderma xylanase*. J. Agric. Food Chem. 50, 1625−1630.

Yu, P., McKinnon, J.J., Maenz, D.D., Racz, V.J., Christensen, D.A., 2002b. The interactive effects of enriched sources of *Aspergillus* ferulic acid esterase and *Trichoderma xylanase* on the quantitative release of hydroxycinnamic acids from oat hulls. Can. J. Anim. Sci. 82, 251−257.

Yu, P., Mcnmon, J.J., Maenz, D.D., Olkowski, A.A., Racz, V.J., Christensen, D.A., 2003. Enzymic release of reducing sugars from oat hulls by cellulase, as influenced by *Aspergillus* ferulic acid esterase and *Trichoderma xylanase*. J. Agric. Food Chem. 51, 218−223.

Yuan, X., Wang, J., Yao, H., 2005. Feruloyl oligosaccharides stimulate the growth of *Bifidobacterium bifidum*. Anaerobe 11, 225−229.

Zaldivar, J., Nielsen, J., Olsson, L., 2001. Fuel ethanol production from lignocellulose: a challenge for metabolic engineering and process integration. Appl. Microbiol. Biotechnol. 56 (1−2), 17−34.

Zheng, L., Zhenga, P., Sun, Z., Bai, Y., Wang, J., Guo, X., 2007. Production of vanillin from waste residue of rice bran oil by *Aspergillus niger* and *Pycnoporus cinnabarinus*. Bioresour. Technol. 98, 1115−1119.

CHAPTER 5

Environmental Impacts and Future Prospects*

5.1 LOOKING INTO THE FUTURE

Forest biorefineries could produce fewer emissions and support sustainable forestry (Miller et al., 2005). The overall environmental implications and life cycle of the forest biorefinery are still being studied. However, there could be a number of positive environmental impacts (Bajpai, 2012). For example, a forest biorefinery utilizing gasification (in a black liquor gasification combined cycle [BLGCC] configuration) rather than a Tomlinson boiler is predicted to produce significantly fewer pollutant emissions due to the intrinsic characteristics of the BLGCC technology. Syngas cleanup conditioning removes a considerable amount of contaminants and gas turbine combustion is more efficient and complete than boiler combustion. There could also be reductions in pollutant emissions and hazardous wastes resulting from cleaner production of chemicals and fuels that are now manufactured using fossil energy resources. In addition, it is generally accepted that production of power, fuels, chemicals, and other products from biomass resources creates a net zero generation of carbon dioxide (a greenhouse gas), as plants are renewable carbon sinks. A key component of the forest biorefinery concept is sustainable forestry. The forest biorefinery concept utilizes advanced technologies to convert sustainable woody biomass to electricity and other valuable products, and would support the sustainable management of forest lands. In addition, the forest biorefinery offers a productive value-added use for renewable resources such as wood thinnings and forestry residues as well as urban wood waste (Mabee et al., 2005).

*Some excerpts taken from Bajpai (2012). Biotechnology for Pulp and Paper Processing with kind permission from Springer Science + Business Media.

Biorefinery in the Pulp and Paper Industry. DOI: http://dx.doi.org/10.1016/B978-0-12-409508-3.00005-5

Black liquor gasification, whether conducted at high or low temperatures, is still superior to the current recovery boiler combustion technology. The thermal efficiency of gasifiers is estimated to be 74% compared to 64% in modern recovery boilers, and the integrated gasification and combined cycle (IGCC) power plant could potentially generate twice the electricity output of recovery boiler power plants given the same amount of fuel (Farmer and Sinquefield, 2003). While the electrical production ratio of conventional recovery boiler power plants is 0.025–0.10 MWe/MWt, the IGCC power plant can produce an estimated 0.20–0.22 MWe/MWt (Farmer and Sinquefield, 2003; Sricharoenchaikul, 2001). This increase in electrical efficiency is significant enough to make pulp and paper mills potential exporters of renewable electric power. Alternatively, pulp mills could become manufacturers of biobased products by becoming biorefineries. Additionally, the new technology could potentially save more than 100 trillion Btu of energy consumption annually, and within 25 years of implementation, it could save up to 360 trillion Btu/year of fossil fuel energy (Larsen et al., 2003). The new technology also offers the benefits of improved pulp yields if alternative pulping chemistries are included, and reductions in solid waste discharges. Also, the process is inherently safer because the gasifier does not contain a bed of char smelt, unlike in recovery boilers, which reduces the risk of deadly smelt-water explosions (Sricharoenchaikul, 2001).

IGCC power plants will reduce wastewater discharges at pulp and paper mills, even though they most likely will not significantly impact water quality (Larsen et al., 2003). Also, IGCC power plants will reduce cooling water and makeup water discharges locally at the mill, and because the efficient gasifiers will cause grid power reductions, substantial reductions in cooling water requirements at central station power plants will also occur (Larsen et al., 2003). Central station power plants have large water requirements for cooling towers in order to provide grid power to customers. Overall, the implementation of IGCC power plants will cause net savings in cooling water requirements and net reductions in wastewater discharges.

The most significant environmental impact caused by black liquor gasification will occur in air emissions. Compared to the current recovery technology, the IGCC system could lower emissions of many pollutants—SO_2, nitrogen oxides (NO_x), CO, VOCs, particulates, and Total

Table 5.1 Relative Emissions Rates of Different Emissions			
Pollutant	Relative Environmental Impact	Relative Emissions Rates with Controls on Recovery Boilers	Relative Emissions Rates with Gasification Technology
SO_2	High	Low	Very low
NO_x	High	Medium	Very low
CO	Low	Medium	Very low
VOCs	High	Low	Very low
Particulates	High	Low–Medium	Very low
CH_4	Low–Medium	Low	Very low
Hazardous air pollutants (HAPs)	Medium–High	Low	Very low
TRS	Low	Low	Very low
Wastewater	Medium–High	Low	Very low–low
Solids	Very low	Low	Low
Source: *Based on Larsen et al. (2003)*.			

reduced sulphur (TRS) gases—and overall reduce CO_2 emissions. Even with improved add-on pollution control features, the recovery boiler system still causes higher overall emissions than the IGCC system (Larsen et al., 1998, 2003). Table 5.1 shows a list of different emissions and their qualitative environmental impact, along with relative emissions rates for both recovery boilers and gasifiers.

Because the biomass sources at pulp and paper mills are sustainably grown, a black liquor gasification-based IGCC plant or biorefinery would transfer smaller amounts of CO_2 to the atmosphere as compared to using fossil fuels. The vast majority of the CO_2 emitted would be captured from the atmosphere for photosynthesis and used for replacement biomass growth, producing O_2 (Larsen et al., 2003). According to Larsen, if the pulp and paper industry converts the 1.6 quads of total biomass energy to electricity, 130 billion kWh/year of electricity could be generated. This electricity generation in a black liquor gasification-based IGCC plant could displace net CO_2 emissions by 35 million tons of carbon per year within 25 years of implementation. Within 25 years of implementation, the IGCC could displace 160,000 net tons of SO_2, as most of the SO_2 produced in the process would be absorbed during H_2S recovery. Moreover, the overall reduction of TRS gases (i.e., H_2S) using gasification technology will also reduce odor, which will improve public acceptance of pulp and paper mills, particularly in populated areas.

The forest biorefinery may represent a viable route to an improved business model for the pulp and paper industry. Indeed, mills have now entered into meaningful dialogue about the possibility for change (Lynch and Chaine, 2007). Cost reductions, mergers and acquisitions, and quality-based innovation have been the strategy in recent years for many pulp and paper companies. In order to implement the forest biorefinery opportunity, it will be essential to refocus, adopt a market-driven mentality, and support ongoing product-centric research and development. An overall change of key success factors (Procter, 2006) is critical, as is modifying the company culture to support new objectives. The key success factors for implementing the forest biorefinery should be premised to a great extent on establishing a viable and sustainable business plan. The business plan definition will be supported by a product design methodology that incorporates market analysis and competitive position determination. This product selection methodology can be used to identify commercial opportunities, quantify potential economic benefits, and determine the best biorefinery product(s) for a given mill that fills market niche and shows promise for the longer term. Supply chain management will be critical in order to maintain high margins over the longer term by optimizing value chain development.

The forest biorefinery offers a business strategy that potential forestry companies are seriously considering for improving the overall financial performance of the sector (Chambost and Stuart, 2007). However, there are considerable technology and business risks related to its implementation. These risks can be mitigated to a great extent by using systematic product and process design tools for analyzing biorefinery strategies. Chambost and Stuart (2007) have described the basis for a systematic product design methodology for rapid market analysis, suitable for evaluating the economic and commercial potential of a biorefinery project, using a set of business tools that includes market and synergy identification. The preliminary application of this methodology has illustrated that the long-term competitive advantage of companies implementing the biorefinery is unlikely to be related to technology, but rather, related to the unique supply chain that companies put in place, coupled with manufacturing flexibility.

Industry leaders, investors, policy-makers, and others are now beginning to better understand the vital role to be played by biorefineries as we move from a fossil fuel-based energy economy toward a biobased one (Connor, 2007). When properly located and operated, the potential of an integrated forest biorefinery is believed to be huge: a very attractive and synergistic business opportunity for both the colocated pulp and paper mill and for the biorefinery itself. Biorefineries are a key pathway to our biofuture, displacing fossil fuels and supplying clean, renewable, and carbon-neutral energy. Biorefineries fit very well at pulp and paper mills because of their inherent ability to gather and process biomass and create energy from biomass. Forest biorefinery has the potential to marry the interests of the society, environment, and the industry (forest sector).

REFERENCES

Bajpai, P., 2012. Biotechnology in Pulp and Paper Processing. Springer-Verlag New York, NY.

Chambost, V., Stuart, P.R., 2007. Selecting the most appropriate products for the forest biorefinery. Ind. Biotechnol. 3 (2), 112−119.

Connor, E., 2007. The integrated forest biorefinery: the pathway to our bio-future. International Chemical Recovery Conference: Efficiency and Energy Management, 29 May−1 June 2007, Quebec City, QC, Canada, pp. 323−327.

Farmer, M.C., 2005. The adaptable integrated biorefinery for existing pulp mills. Presentation at TAPPI Engineering, Pulping, and Environmental Conference, 28−31 August, Philadelphia, PA.

Farmer, M., Sinquefield, S., 2003. An External Benefits Study of Black Liquor Gasification. Georgia Institute of Technology (Final Report, 15 June 2003).

Larsen, E., Kreutz, T., Consonni, S., 1998. Performance and preliminary economics of black liquor combined cycles for a range of Kraft pulp mill sizes. International Chemical Recovery Conference, vol. 2, 1−4 June 1998, Tampa, FL, pp. 675−692.

Larsen, E., Consonni, S., Katofsky, R., 2003. A Cost-Benefit Assessment of Biomass Gasification Power Generation in the Pulp and Paper Industry. Princeton Environmental Institute (Final Report, 8 October 2003).

Lynch, H., Chaine, B., 2007. Capital expenditures: the cost of business. Pulp Pap. Canada 108 (4), 14−16.

Mabee, W.E., Gregg, D.J., Saddler, J.N., 2005. Assessing the emerging biorefinery sector in Canada. Appl. Biochem. Biotechnol. 121−124, 765−777.

Miller, M., Justiniano, M., McQueen, S., 2005. Energy and environmental profile of the U.S. pulp and paper industry. Prepared by Energetics Incorporated, December 2005, Columbia, MD.

Procter, A., 2006. Key success factors: a guide for prioritizing performance improvement effort. Pulp Pap. Canada 107 (7−8), 59.

Sricharoenchaikul, V., 2001. Fate of Carbon-Containing Compounds from Gasification of Kraft Black Liquor with Subsequent Catalytic Conditioning of Condensable Organics. Georgia Institute of Technology, Atlanta, GA (Ph.D. Dissertation).